D1251391

Green Thought in German Culture

Green Thought in German Culture

HISTORICAL AND CONTEMPORARY PERSPECTIVES

Edited by

Colin Riordan

UNIVERSITY OF WALES PRESS • CARDIFF • 1997

© The Contributors, 1997

British Library Cataloguing-in-Publication Data.
A catalogue record for this book is available from
the British Library.

ISDN 0–7083–1421–X

Typeset by Action Typesetting Ltd, Gloucester
Printed in Great Britain by Dinefwr Press, Llandybïe

Contents

Acknowledgements

My colleague Tom Cheesman took an equal part in the conception, planning and early editing process of this book. I am grateful to him for his support, and particularly for his translation of chapter 16.

Preface

Thinking green is now part of the very fabric of life in Germany. Political discourse is inconceivable without at least a genuflection in the direction of the environment. People's everyday lives have been altered to include green disposal and recycling as a matter of course, and habits of consumption are informed by considerations which would have been unknown a generation ago. This shift in attitudes is evidence of cultural change in the broadest sense. But everyday environmentalism is only the visible evidence of a more profound, enduring and well-established cultural phenomenon. Donald Worster has found that the science of ecology is, perhaps more than mathematics or thermodynamics, the product of 'specific cultural conditions'.[1] In Germany, green ideas have a peculiar and complex resonance: their historical associations are diverse and splintered, ranging from Romantic nature philosophy to anthroposophy, from anarcho-socialism to fascism, from heady nationalism to scientific apocalypse. A historical and contemporary consideration of the influence of green thought on German culture is, then, a product of the recognition that green thought is indivisible from German culture. This volume considers the emergence and development of green ideas in Germany, and analyses to varying degrees the part played by green thought in literature, drama, film, popular fiction, art, architecture, philosophy and other cultural arenas. In so doing, it explores the formative role that writers, artists and other cultural actors have played in disseminating, elaborating and criticizing green ideas.

Part I ('Historical Perspectives') opens with an introduction to the origins and development of green thought in Germany from the early nineteenth century onwards. Green ideas are broadly divided into the instrumental, managerial strand of environmentalist concern, protest and activism on the one hand, and the philosophically and ideologically grounded category of ecologism, of which environmental concern is only a part, on the other. A sceptical stance is taken towards the question of how far green ideas can be said to have existed in any meaningful modern sense in the nineteenth century, and the resonances of the German historical legacy for modern green politics are assessed. The subsequent three contributions consider in turn the association between early green thought and bourgeois conservatism, ecosocialism and Nazism. Matthew Jefferies places the origins of the Wilhelmine environmental movement in a wider cultural-reform tendency, whose concerns were primarily aesthetic in nature. This view is illustrated using the example of the architect, cultural commentator and conservationist Paul Schultze-Naumburg, whose views on the relationship of the built environment to the natural world led him to explore some of the unforeseen ecological effects of human interference in nature. If Schultze-Naumburg is a model representative of the nationalist (and ultimately fascist) strand of German conservationism, then Gustav Landauer, subject of the next chapter, exemplifies early ecosocialism. Williams argues that Landauer's anti-industrialist, communalist, anti-Marxist critique is paralleled in Worringer's theories of art and is mediated through the work of Expressionist writers. Part I closes with an example of the way in which Nazi appropriation of some elements of the ecological idea provoked complex literary and artistic reaction. Brian Keith-Smith argues that Lothar Schreyer moved from embracing 'blood and soil' ideology, through celebrating the affinity of 'Nordic man' with the landscape and forest, to a vision of purified man in harmony with nature which ultimately flowed into a Christian ecological understanding of nature.

In chapter 5, which opens part II ('Contemporary Trends'), Ingolfur Blühdorn considers the importance of the Frankfurt School of critical theory for an understanding and resolution of the crisis which he diagnoses in the current eco-debate. He elaborates Adorno's conceptualization of 'nature' as a social and historical phenomenon ('second nature'), arguing that by adopting this concept, green thought will be able to advance belatedly from a modernist to a postmodernist phase, characterized by the abandonment of faith in the progressive value of universal rationality. By contrast, Thompson offers a materialist analysis of the perception of ecological crises, posing the provocative question: 'How can ecology be rescued from the ecologists and turned back into a central part of a wider social agenda?' He takes the work of the prominent green thinker Rudolf Bahro as a prime example of the intrinsically irrational and anti-humanist nature of contemporary green thought, which eventually led Bahro to advocate a theocratic 'eco-dictatorship'. In chapter 7, Statham also addresses the question of whether green thought inherently tends towards an authoritarian politics in her examination of the way in which the intellectual New

Right in Germany has adopted elements of green discourse in support of an extreme nationalist and conservative platform. She considers the continuity of these contemporary trends with early twentieth-century green ideology on the far right, finding many similarities but also crucial differences. The final chapter in this section might have been equally at home in the next, but its analysis of Irmtraud Morgner's remarkable anticipation of the 'ecofeminist' debate in the USA, Britain and West Germany raises questions both about the validity of ecofeminist theory, and about the role of literature as a vehicle for theory and for the critique of theory. Morgner's *Amanda*, published in the German Democratic Republic (GDR) in 1983, reworks classical myth and legend (Pandora, the Sirens) in an exploration of gender inequality as the root cause of ecological crisis.

Part III ('Post-War German Literature') begins with a broad survey of catastrophist literature in the post-war period, concentrating on Enzensberger, Müller, Kunert, Horx, Pausewang and, especially, Grass and Wolf. The chapter distinguishes mythical from ecological apocalypse, and includes a summary of the history of apocalyptic writing in German literature. Goodbody concludes that the aesthetically most satisfying works in the catastrophist genre are often ambiguous as ecological statements, and paradoxically tend to subvert doomsday prediction. Günter Grass, as the most prominent post-war German writer, and one whose interest in matters green goes back to the early 1970s, deserves special attention in this volume. His vision of global catastrophe in *Die Rättin* [*The Rat*] (1985) contemplates questions of ecological destruction with rare sophistication. Siemon shows how green ideas are incorporated into the very narratological assumptions which underlie Grass's novel. In the course of his analysis he considers the end of human history, the disappearance of the narrator and the end of narration. In conclusion, Siemon is able to show that Grass's method implies hope in the face of catastrophe: even as the subject and sense of his story disappear, it is possible to detect a refusal to accept the postulated annihilation as inevitable. Moving to a gentler tradition, in chapter 11 Bill Niven identifies a counterpart to the traditional eighteenth- and nineteenth-century *Bildungsroman* in three post-war novels. While the traditional *Bildungsroman* shows characters learning to adapt to social norms, the heroes of post-war green *Bildungsromane* begin by conforming to the prevailing ethos of economic progress, but gradually learn to discard those values in favour of environmental sustainability. After a brief survey of precursors including Heine and Döblin, Niven uses works by Frisch, Timm and Cramer to consider how the gradual emergence of ecological consciousness is represented. Niven concludes with an examination of the subtext of these works, posing the question of how successful their supposed transformation really is: do these works support or subvert the proposition that ecological destruction can be stopped? In the next chapter, Martin Kane considers the treatment of environmental issues in GDR literature in the light of revelations on the environment which have emerged since the collapse of the GDR, concentrating particularly on the way in which uranium mining was presented. The tension between belief in the liberating power of technology to transform socialist society (a belief which persisted until the 1980s) on the one hand, and the dangers to health and environment which such exploitation engenders on the other, informs Kane's analysis of novels by Martin Viertel and Werner Bräunig. The chapter includes a brief survey of the way ecological issues – largely a taboo subject – were treated by GDR writers throughout the existence of the state. In chapter 13, Jürgen Barkhoff identifies a specifically Swiss form of ecological literature, in which a pre-modern conception of nature conditions depictions of the world as a living being ('Gaia'), which defends itself against the depredations of civilization by destroying the order imposed on it by humankind. The close proximity in Switzerland of powerful symbols of the natural world (mountains, glaciers) and an isolationist, regulatory society which encourages conformity highlights the increasingly irreconcilable contradictions between the two. Barkhoff considers both prominent writers such as Frisch and less well-known figures such as Hohler.

In part IV ('Art and Popular Culture') the role of green thought beyond literature in contemporary German culture is explored, with particular emphasis on cultural artefacts as a means of influencing, as well as representing, popular ecological consciousness. In chapter 14, Frank Finlay examines ways in which green thought as part of the wider intellectual tradition informs the aesthetic approach of Joseph Beuys, who was a founder member of the German Greens. Taking account of the artist's collaboration with Böll, and of the influence on him exerted by the Frankfurt School, Romanticism, Schiller's idealism, the Expressionist sculptor

Lehmbruck and Steiner's anthroposophy, Finlay considers Beuys's attempt to instrumentalize art as a means of empowering individuals to fulfil their human potential and recover a state of harmony with the natural world. He concludes with an exploration of the complex tensions which led to Beuys's disenchantment with Die Grünen. Eco-drama is the topic under discussion in chapter 15, in the form of *Naturtheater*; non-professional theatre performed in open-air settings which advertises itself as 'close to nature' and corresponds formally in key respects to the community-based, anti-élitist culture advocated by many green thinkers. The author, director and performer Martin Schleker, of the Hayinger theatre, uses the *Naturtheater* as a vehicle for ecologically orientated social critiques. Alison Phipps examines his work and its local and critical reception, drawing on her experiences as participant observer with the theatre's cast and crew, and considers the extent to which this form of theatre may be especially well placed to raise popular ecological consciousness. Birgitta Schüller poses a similar question with respect to crime fiction in Germany. The *Öko-Krimi* (crime fiction dealing with ecological issues) has arguably emerged as a distinctive German subgenre. Schüller discusses some work which makes merely decorative use of 'green' themes, treats them superficially, and/or is of the lowest literary quality. But she also shows that some writers – notably the second-generation Turkish author Jakob Arjouni – are effectively harnessing the generic forms of popular fiction to raise ecological consciousness, and are experimenting with these forms in ways which at least hold out the promise of interesting future developments. In the final chapter of the volume, Tom Cheesman attempts a defence of the film-maker Werner Herzog against the charge of being an eco-criminal both intellectually and literally: the making of *Fitzcarraldo* not only hastened the destruction of the environment and indigenous culture of its location, but it also cost several locals' lives. Herzog's appropriation of the apocalyptic tradition, his critics maintain, amounts to little more than an aestheticization, and thus a validation of catastrophe. In fact, Cheesman argues, Herzog's apocalyptic imagination is in the popular, critical tradition which maintains hope in the face of extreme suffering.

It will be clear from the above outline that *Green Thought in German Culture* is a deliberately all-embracing title. The volume is meant to be wide-ranging and eclectic. In that spirit, 'green thought' is taken to mean any ideas, philosophies or ideologies which are concerned with ecology or the environment. Similarly, 'culture' is an inclusive term from which political theory and activism cannot be abstracted. And although the main focus of the volume is on Germany, 'German' is taken to include Austria, Switzerland and the GDR as well as the present-day Federal Republic. Given its broad range of subject matter, the volume makes no claim to be comprehensive. But it does hope both to reconsider familiar matters from a green perspective and – taking its cue from the rich diversity of green thought – to illuminate some of the less frequented areas of German culture.

Notes

[1.] Donald Worster, *Nature's Economy: A History of Ecological Ideas*, 2nd edn. (Cambridge, 1995), xi.

Notes on Contributors

Colin Riordan is Senior Lecturer in German at the University of Wales, Swansea. He has edited books on German writers and the Cold War and on Peter Schneider, and is the author of a book and several articles on Uwe Johnson.

Jürgen Barkhoff is Lecturer in German at Trinity College Dublin. He is the author of a book on Mesmerism and Romantic literature (1995) and co-editor (with Eda Sagarra) of a volume on anthropology and literature around 1800 (1992). He has published articles on Goethe, Herder, exile literature and Max Frisch.

Ingolfur Blühdorn is Lecturer in Contemporary German Studies at the University of Bath. His area of specialism is social-movement research, and in particular contemporary ecological theory. He has published several articles and co-edited *The Green Agenda: Environmental Politics and Policies in Germany* (1995).

Tom Cheesman is Lecturer in German at the University of Wales, Swansea. He is the author of *The Shocking Ballad Picture Show: German Popular Literature and Cultural History* (1994) and several articles on folklore, literature and cultural studies. He is currently working on a study of *The Goethe Ballads*.

Frank Finlay is Senior Lecturer in German in the Department of Modern Languages at the University of Bradford. He is author of a monograph and several articles on Heinrich Böll.

Axel Goodbody is Senior Lecturer in German Studies at the University of Bath. He has published widely on twentieth-century poetry (Eich, Enzensberger, Fried, Kramer, Lehmann) and GDR literature. He has recently published a reader of environmental literature (*Umwelt-Lesebuch*, Manchester University Press), and is currently editing a volume of critical essays entitled *Literatur und Ökologie* (Amsterdamer Beiträge).

Matthew Jefferies is Lecturer in German History at the University of Manchester. His particular interest is the cultural history of Imperial Germany. His publications include the 1995 book *Politics and Culture in Wilhelmine Germany: The Case of Industrial Architecture*.

Martin Kane is Reader in Modern German Literature at the University of Kent at Canterbury. He is the author of *Weimar Germany and the Limits of Political Art* (1987), the editor of

Socialism and the Literary Imagination: Essays on East German Writers (1991), and has published several articles on contemporary German writing.

Brian Keith-Smith retired early as Reader in German, University of Bristol. He has published and edited books on Bobrowski, Büchner, Expressionism, Austrian writers, women writers and Schreyer. He also lectures on art history for the National Association of Decorative and Fine Arts Societies.

Bill Niven is Lecturer in German at the University of Aberdeen. He has published a book on the reception of Friedrich Hebbel during the Third Reich, as well as a number of articles on National Socialist literature. He is also interested in post-1945 German literature, and has published several articles on Christoph Hein. Currently he is writing a study of Christoph Hein's *Drachenblut*.

Alison Phipps is Lecturer in German at Glasgow University. She completed a Ph.D. on *Naturtheater* in 1995, with the generous support of the Leverhulme Trust, and has interests in performance research, contemporary German cultural studies, ethnography and eco-theology.

Antje Ricken is a doctoral candidate in the Department of German at the University of Wales, Swansea. Her research interests centre on post-war women's fiction, and on Irmtraud Morgner in particular.

Birgitta Schüller teaches in the Göttenbach-Gymnasium, Idar-Oberstein. Her research interests include migrant literature, contemporary German literature, German as a foreign language, intercultural communication and didactics.

Johann Siemon teaches in a German grammar school. He is the author of several articles on Gottfried Benn, Uwe Johnson and Carl Einstein.

Alison Statham is Lecturer in Contemporary German Studies at the De Montfort University, Leicester. Her Ph.D. thesis examines the development of extreme right-wing cultural thought in Germany. Particular emphasis is given to the post-1945 ideologies of conservative ecology and Euro-scepticism.

Peter Thompson is Lecturer in German at the University of Sheffield. His primary research and teaching areas are the politics and history of post-war Germany, and he has also published on the philosophy of history and the political impetus behind receptions of Brecht.

Rhys W. Williams is Professor of German at the University of

Wales, Swansea. He has published extensively on German Expressionism (Sternheim, Benn, Einstein, Kaiser and Toller) and on contemporary literature (Andersch, Böll, Lenz and Walser).

Part I
Historical Perspectives

1 • Green Ideas in Germany: A Historical Survey

Colin Riordan

Historians often draw attention to the ahistoricism of green debate. The deficiency is typically ascribed to the future-orientated motivation of green activism, its ephemeral and rapidly changing nature, or to a desire to suppress or de-emphasize uncomfortable elements of green history.[1] This is not to say that no green history has been written: since 1984 there have been many responses to Ulrich Linse's complaint 'against the ahistoricism of ecological thought', at least so far as Germany is concerned.[2] Yet the problem persists, for the legacy of German history is a complex one. The continued success of the German Greens (despite pessimism about their future)[3] is not merely a matter of a favourable electoral system and a history of post-war radicalism. Green ideas have a peculiar and powerful resonance in Germany, yet are burdened with unwelcome associations largely absent in the USA or Britain. This chapter will attempt to show the way in which green ideas have gradually been woven into the German cultural fabric during the last two hundred years or so, and will assess the importance of historical precedents for the modern German green movement.

The immediate question to arise is what we mean by green ideas. The first distinction commonly drawn is between anthropocentric (human-centred) and non-anthropocentric approaches.[4] Joachim Radkau has questioned whether this distinction is genuinely useful, since 'a history of "nature in itself" cannot be written, only the history of a nature that has already somehow been defined by human vested interests.'[5] The recognition that nature is a fluid social construct rather than some static separate entity has transformed green debate. Moreover, such terms seem to favour existential categories above political ones, and since green ideas aim to change the human condition, they are by their essence political. With this in mind, a more useful distinction might be one which is repeatedly stressed by Andrew Dobson in his *Green Political Thought*: that between environmentalism and ecologism.[6] Environmentalists, in this view, are concerned to mitigate the worst effects of industrialism on the environment without fundamentally challenging the prevailing orthodoxy. They adopt a pragmatic or managerial approach, campaigning on particular fronts, or against new developments which threaten the natural world. Ecologists, on the other hand, believe that tinkering with or restraining industrial society is wholly inadequate. The finite nature of the earth's resources requires nothing less than a radical change in the structure of society, change which arises from the abandonment of economic growth and big industrialism in favour of small-scale, dispersed economies and egalitarian democracy.[7] The distinction is familiar to anybody acquainted with green politics in Germany, subject during the 1980s to disputes between environmentalist *Realos* (realists) and ecologist *Fundis* (fundamentalists). Surprisingly, it is not one commonly made in histories of green ideas in Germany. Yet historically, that distinction is important. While it can be shown that environmentalism has a relatively long history in Germany, ecologism is a more recent phenomenon.[8] That is not to say that the origins of ecologism are necessarily recent. But before looking at origins, we need to decide how the two categories are to be recognized.

Rhetorical objections to human interference in the natural environment (such as deforestation or river channelling in the nineteenth century) might indicate a hitherto unsuspected level of early ecological awareness, but are not in themselves evidence of environmentalism in a modern sense. Environmentalism proper can reasonably be said to exist where there is evidence of active protest or campaign against damage to the natural environment which rests on a knowledge of the ecological effects of human action. Where concern for the preservation of a particular natural feature is actuated primarily by aesthetic considerations, however, it makes more sense to speak of conservationism, the history of which has been extensively documented in Germany.[9] This distinction is not always easy to make, as will become clear later in this chapter. Broadly, conservationists were concerned to preserve some aspect of the landscape, or even of the built environment, from the effects of industrial development. Whether their concerns were in the main aesthetic or anti-modernist, rather than strictly environmental, is a matter for debate; most experts incline towards the former view.

On their own, environmentalist objections to pollution, depletion of the water table and so on do not amount to ecologism. Ecologism is an ideology composed of a number of elements. Some of the elements have an ancient lineage, but together they amount to a coherent world-view which, Dobson suggests, cannot have existed much before the publication of *Limits to Growth* in 1972.[10] The ecologist world-view is holistic, that is, all things both organic and inorganic are interrelated and interdependent; interfering with one will have unpredictable effects on others. Ecologism must be based on scientific understanding of the global effects of human interference in nature, but is characterized by a critical stance towards the effects of science (in its technological guise) on global ecology. Human beings are a part of nature, but have no superior status or special right to exist within it. The prescription for radical change outlined above is accompanied by apocalyptic warnings of global doom, at least for the human race, in the event of inaction. At the risk of some simplification, ecologism is holistic, critically scientific, ecocentric, prescriptive and apocalyptic. There are varying admixtures of misanthropism and mysticism.

Curiously, as I mentioned earlier, although ecologism is the more recent phenomenon, some of its elements are far older than environmentalism. Heraclitus (535–475 BC) speaks of the harmony of the cosmos, in which nature operates according to immutable laws. The natural world is in eternal flux; human beings are a mere part of the order of things, subject to a fate ordained by nature.[11] We need look no further than Plato for concepts of the unity of humankind and nature, of reason and cosmos.[12] And there is abundant evidence of early human cultures such as Native American peoples who believed in a harmony of humankind and nature which depended on the mutual interdependence of plants, animals, landscape and people. An awareness of sustainability went hand in hand with such beliefs. Holism thus has a long history in human culture, and its reverberations may be detected in the German tradition as elsewhere. Even as the evolving scientific method was engendering the atomistic view of the universe which worked counter to holistic world-views, Jakob Böhme (1575–1624) was proposing a kind of christological pantheism, in which the whole of creation is a revelation of the omnipresent, unfathomable God, in which all things have been created into one being.[13] The attempts of Leibniz to demonstrate the unity between the endless diversity and the wholeness of nature through a view of the universe which combines its all-embracing harmony with its monadistic nature led Mayer-Tasch to dub Leibniz one of the fathers of ecological thought.[14] It is important to remember, however, that such evidence of holistic thought is by no means evidence of ecologism; merely of one of its constituent, if essential, parts.

The same can be said of Rousseauistic notions of natural harmony. Rousseau's ascription of greater nobility to the savage living in harmony with nature over civilized man alienated from his natural origins certainly has resonances for the ecological idea. And Rousseau's ideals clearly influenced those eighteenth-century German writers for whom the destiny and happiness of human beings was to be found in recreating our relationship with nature.[15] Yet this recourse to a supposed golden age of natural harmony is still a far cry from later understandings of the impact of man on nature, and the ultimate consequences of their conflict. Similarly, other detached elements of the ecological idea may be detected in German culture towards the end of the eighteenth century.

It is not only Goethe's pantheism which leads him to be cited frequently as a forerunner of ecological thinkers,[16] but also his efforts to encompass the totality of human endeavour in all areas. What sets Goethe apart in this respect is his attitude to science, and, in particular, to the relationship between science and nature. Ecologism is beset with the paradox that it simultaneously needs science in order to provide a sound basis for a critique of the effect of industrialism on the environment, but wishes to reject, restrict or, at best, radically reform science in its technological manifestation precisely because of the destructive effects of technology. For this reason some Greens have resorted to mystical explanations of the oneness of nature, and to Gaian myths that have almost universal provenance.[17] As a natural scientist some of whose results remain valid today, Goethe was far from rejecting science itself. He did, however, emphatically reject the scientific method. The reasons for this emerged in his celebrated quarrel with Newton, whom he attacked for reducing nature to an object for dissection and analysis. His disagreement with Newton over the composition of light, which Goethe regarded as a unity of white, in contrast to the Englishman's theory of spectral analysis, is symptomatic of his view of nature.[18] For Goethe, nature was not an object to be analysed, but part of a seamless web which encompassed human life as well as the natural environment. Nature was to be perceived not only with rational but also with aesthetic tools; indeed, the two were, for Goethe, epistemologically inseparable.

Goethe's passionate defence of nature against analytical Newtonian physics was shared by the German Romantics. The primacy of the organic over the mechanical was axiomatic to the Romantic movement, but a simultaneous fascination with and suspicion of technology was prevalent. The dangers which can arise when people attempt to interfere in natural processes, setting off sequences of events which rapidly spiral out of control, are commonly represented in the Romantic literature of the period. One need only think of Mary Shelley's *Frankenstein*, or in Germany, of E. T. A. Hoffmann's *Der Sandmann* [*The Sandman*], in which a life-sized clockwork doll deceives a young man into falling in love, eventually driving him to madness and suicide. There is a powerful sense of ambivalence about the tale – it is never clear whether the danger is psychological, mystical or technological – but the very ambivalence concerning the benefits and dangers of technology has strong resonances in modern ecologism. Moreover, the nature philosophers of the Romantic period conceived of nature as 'an organism, not divisible into parts', a concept echoed in the modern ecological understanding of nature, and of man as being 'originally endowed with an "inner" or "universal" sense which *knew* the universe . . . this sense still exists in us, though smothered, and all we have to do is to descend to it in order to regain attunement and integration, to return to our original harmony with nature'.[19]

It is clear, then, that elements of ecological thought have been present in German culture at least from the eighteenth century onwards, and that some aspects are part of a much longer and broader cultural history. Myths of apocalypse (as distinct from scientific apocalypse) have origins reaching back at least to the Old Testament.[20] The scientific basis of ecologism – the relationship between organisms and their environment – can be traced to Linnaeus, and was first explicitly formulated in Germany in the latter half of the nineteenth century by the biologist Ernst Haeckel.[21] The other essential elements accrued gradually. The notion that radical social change is necessary gathered momentum from the end of the nineteenth century, reaching an early peak in Germany during the Weimar period. At this time also, the recognition that resources (especially fossil energy resources) are finite began to become prevalent. The prospect of ecological apocalypse gathered credence, and, indeed, potential reality in the 1950s. During that decade enough scientific expertise in the detection of ubiquitous traces of chemical and radioactive pollution was acquired to underlie a reasoned case for the global consequences of unchecked industrialism. It is from this point onward that proto-ecologism can be said to exist in Germany. The first stirrings of conservationism, by contrast, may be identified at the beginning of the nineteenth century.

The seeds of conservationism were sown by the geographer and explorer Alexander von Humboldt, who in 1799 coined the term *Naturdenkmal* [natural monument] to denote a natural feature worth preserving. More than this, Humboldt was, as Worster has put it, 'a pioneer in ecological biology',[22] part of whose *Ansichten der Natur* [*Views of Nature*] (1808) was intended as 'a vivid portrayal of organisms (animals and plants) in their topographical, that is local relationship to the many-faceted surface of the earth (as a small part of the whole of life on earth)'.[23] Many of Humboldt's insights remain valid today: he realized, for example, that a given plot of land yields more food when given over to plant cultivation than when devoted to animal grazing.[24] Interestingly, Percy Bysshe Shelley was to make the same observation in a tract on vegetarianism published some eight years later.[25] Humboldt's investigations into the 'harmony of nature' (inspired in part by conversations with his brother Wilhelm and with Goethe) led him to recognize the dangers of human interference in the environment.[26] He explained the devastating effects of deforestation on the water table in Latin America and the Mediterranean region, for example.[27] The German forest itself, however, had long been supposed to be on the brink of disaster. According to Radkau, countless fears were expressed from the sixteenth to the nineteenth century concerning the threat of timber scarcity and 'ruin of the woodlands'. This should not necessarily be viewed as evidence of ecological crisis; all too often, it was in the state's interest to propagate a myth of forestry mismanagement in order to legitimate state control of timber resources.[28] Some contemporary commentators do, nevertheless, display a remarkably advanced understanding of the dangers. Among the most articulate was Ernst Moritz Arndt.

Arndt (1769–1860) was an early advocate of German unification, a professor at the University of Greifswald who used his journal *Der Wächter* [*The Watchman*] (1815–16) to give vent to his patriotic fervour and frustration over the Congress of Vienna. So far as the environment is concerned, Arndt is notable for being among the first in a long tradition in Germany to make a connection between nationalism and the natural environment. As we shall see, during the course of the nineteenth century, many defenders of the German landscape against the depredations of industrialism were actuated by a desire to protect the spirit of the German *Volk*, which was thought to reside in the rivers, hills and, above all, the forests of Germany. While this bears little relationship to ecologism in a modern sense, such conservationist instincts did, in the case of Arndt, lead to the articulation of some advanced ideas:

> In the last twenty or thirty years, we have seen noble trees and forests in many tracts of German landscape subjected to the most unholy and dastardly mischief. Whole forests have been cut down and whole areas denuded because the individual owner can do as he pleases with nature in the most arbitrary way. What difference does it make to somebody who needs money and intends to use up in ten years something that was meant to sustain his great-grandchildren, if he leaves behind a barren tract of land which will in future yield little pleasure and be of barely any use to people?[29]

These words, written in 1815, reveal not only a concern about deforestation, but also an apparent awareness of the problem of sustainability. The notion that future generations will suffer for misuse of natural resources in the present was not unique at this time, but it was certainly unusual.[30] Moreover, the implication that private ownership of land may be detrimental to the natural environment identified a dilemma which was to plague nationalistic conservationism through to the twentieth century. Arndt's concern for the forest, however, is part of a long tradition which held the German forest to embody the very essence of Germanness.[31] The association between protecting the German forest and protecting the German *Volk* was to remain powerful until the end of Nazism, and exists in extreme remnants today. It was given greater impetus by an intellectual precursor of later conservationists, Wilhelm Heinrich Riehl.

Born in 1823, Riehl was a pioneer in the emerging discipline of ethnology; his major work *Naturgeschichte des Volkes als Grundlage einer deutschen Sozial-Politik* [*Natural History of the People as the Basis of a German Social Policy*] (1851–5) dominated the subject for decades.[32] His *Naturgeschichte* amounts to an attempt to prove that the essence of the German *Volk*, and its regional variations, could be explained by reference to the landscape and climate of Germany. Industrialism and the expansion of the transportation network, in affecting landscape and air quality, affected Germanness itself.[33]

In *Land und Leute*, the first volume of *Naturgeschichte*, Riehl devotes many closely argued pages to showing that the moral, political and economic health of the German nation depends upon the forest, and that a proper balance between agriculture and forestry must be maintained. Riehl's conservatism is romantic and medievalist. He has no brief for vested commercial interests; indeed, he repeatedly stresses that the common ownership of forest is vital to any nation's moral and economic prospects: 'The future of the great Slav empire ... is guaranteed in Russia's impenetrable forests, while an already half-way exhausted national character [*Volkstum*] peers out from the English and French provinces, which no longer have genuine woodlands.'[34] He goes on to suggest that the North American wilderness forms the common stock capital of the United States, and warns against squandering the asset.[35] Riehl does not identify industry as the sole enemy, although the greed for profit and the expansion of railways are both excoriated, but rather revolutionaries who wish to replace forest with agricultural land in order to feed the people of Europe. Anticipating a later ecologist *sine qua non*, he predicts catastrophe unless there is change. Riehl's apocalyptic vision is not of smoking chimney stacks covering the country, but of unremitting agriculture:

> For any natural person, there is something horribly sinister about the thought of seeing every piece of earth ploughed up by human hands; but it is particularly distasteful to the Germanic mind. Were that to happen, then the Day of Judgement would surely be at hand.[36]

Nor is Riehl's fear confined to the forests; he includes other areas of 'Wildnis' [wilderness][37] such as dunes, moors and heaths, mountains and glaciers among the areas which should be celebrated and preserved. Riehl's argument, while in some measure absurd to modern ears, is passionate, vehement, and evangelical. Though not part of an organized campaign, it certainly has qualities of revolutionary fervour which would not be out of place in environmental campaigning. Yet Riehl was not actuated solely by romantic anti-industrialism, writing as he was in the years before industrial evolution became revolution in Germany. In fact, the argument is primarily political. Conservationism in Riehl is a political and cultural matter; in Germany in the mid-nineteenth century, the environment, or at least the preservation of 'wilderness', was already a political issue. In this sense, one might trace the intellectual origins of German environmentalism to this period.

The German roots of modern ecologism were also being formed in the years following the publication of Riehl's *Naturgeschichte*, though independently both of that publication and of conservationism. The Darwinist Ernst Haeckel, Professor of Zoology and Comparative Anatomy at the University of Jena from 1862 to 1909, famously coined the word 'ecology' in 1866,[38] and had an extraordinary impact worldwide. His importance lies not only in his popularization of Darwin (whose *Origin of Species* appeared in German in 1860), but in the way in which he combined science with Goethean spiritual holism. Although the resulting philosophy, which he dubbed monism, had more in common with a nature religion than with science, it was Haeckel who made the crucial connection between scientific and spiritual holism. Monism held that all things in nature, whether physical or spiritual, were one, and that the doctrine of evolution determined all aspects of being.[39] As Daniel Gasman has shown, the social Darwinism which must of necessity arise from such beliefs was more than mere analogy. Monist thinking accepted a 'literal continuity between the laws of nature and the laws of society . . . Just as man was a product of nature so too was the society in which he lived a direct outgrowth of the natural world. Neither history nor its institutions represented a break or departure from nature in any way.'[40] The continuity of man and nature led Haeckel to a rejection of the whole western humanist tradition, and in particular of anthropocentrism: 'As our mother earth is a mere speck in the sunbeam in the illimitable universe, so man himself is but a tiny grain of protoplasm in the perishable framework of organic nature.'[41] While this rejection of human-centred thinking can hardly be termed ecocentric, it is certainly part of a biocentric philosophy. And although anthropocentrism had been under attack since the seventeenth century, this was non-anthropocentrism with a scientific basis, however leavened by mysticism Haeckel's monism may have been. Haeckel's views were propagated through the German Monist League, which was founded in 1906 and whose members included prominent scientists such as Wilhelm Ostwald (the League's president), and Wilhelm Bölsche, originator of 'erotic monism', whose extraordinary popularity should not be underestimated.[42] Unlike modern ecologism, monism was only indirectly concerned with environmentalism, but the primacy of nature was unmistakable. Conservation and environmental movements began to prosper with the popularity of this bond of science, nature and philosophy, together with the nationalism which was building to a pre-1914 peak.

Commentators often rightly note the complexity and obscurity of the history of early environmental protest in Germany.[43] There are records of protests against individual incidences of pollution and desecration of 'natural monuments' reaching as far back as the early eighteenth century.[44] But the question of how far such protests are environmental is a matter for debate. One of the most often-cited cases was the public objection to quarrying the Drachenfels on the Rhine in the 1820s. Locals successfully prevented further damage to the ruins and the rock outcrop on which they stand. Sieferle has rightly questioned whether this is a genuine case of *Naturschutz* [protecting the natural environment], or merely a concern for tradition.[45] From the mid-nineteenth century onwards, however, there was sufficient empirical scientific knowledge available to permit informed objections to the environmental effects of industrialization, even if some of the effects were hotly disputed by scientists themselves.

While categories inevitably lead one to overlook subtle shadings of differentiation, it will suit the present purpose to divide early environmental protests and conservation movements into three: protests and measures against industrial pollution; nature conservation groups and organizations; and alternative life-style groupings.

The deleterious effects of industrial emissions both on the natural environment and on public health were scientifically recognized at the latest by the mid-nineteenth century, though protests were isolated and made in response to specific instances of pollution. Localized protest against air pollution from a nearby ironworks is recorded in Saxony in 1846, when the agricultural chemist Adolph Stöckhardt confirmed that emissions were damaging local vegetation. In 1850 he showed that, given contemporary methods of measurement, there was no safe limit for sulphur dioxide emission so far as trees were concerned. The reaction of authorities was to build higher chimneys, and sometimes to install desulphurization plants. It was apparent even to contemporaries, however, that the needs of industry took priority.[46] Nor is there evidence that environmental protection *per se* was the aim in taking such measures; the primary concerns were commercial (manufacturers were forced to pay compensation for damaged forests) and public-health-orientated. Public health was also the main concern so far as water pollution was concerned. Before the 1860s in Germany, sewage was disposed of in the traditional manner: haphazard use of local streams, burial, or recycling as fertilizer. Rapid urbanization caused large conurbations to follow the English lead in establishing sewage systems which emptied into rivers. The concentration of sewage in this way led Prussia in 1877 effectively to forbid the use of rivers for this purpose; the affected conurbations lobbied to have the ruling reversed. There were also objections to industrial pollution; in 1877, 140 complaints about river pollution were made in Saxony, of which 93 per cent attributed the cause to industry.[47] In response, opponents formed the Internationaler Verein gegen Verunreinigung der Flüsse, des Bodens und der Luft [International Union against the Pollution of the Rivers, Soil and Air] in Cologne in October 1877. At the heart of the dispute was scientific disagreement over the ability of rivers to neutralize pollutants, and a direct conflict between human public-health needs and those of the natural environment. After some initial success for the Association, by 1888 permitted levels of sewage outflow were raised, so that, effectively, few restrictions remained.[48] Throughout, commercial considerations and public health in towns took priority.[49] This may well be understandable, but although there is evidence here of recognizably environmentalist concerns (a knowledge of ecological damage and protest against it) those concerns remained peripheral to the main argument.

The extremely rapid rate of catch-up industrialization in Germany after the 1860s provoked a twofold conservationist response: government-sponsored nature-conservation agencies set up by the German states, and public associations formed to further the cause of a specific form of nature conservation. These organizations were conservationist rather than environmentalist, although as scientific knowledge improved, some of them evolved into environmental campaigning groups. Prussia, Bavaria and Württemberg all set up conservation agencies around the turn of the century in response to the mass nature tourism caused by urbanization and the expansion of rail.[50] The German Garden City movement, which originated near the beginning of the nineteenth century, was reorganized in 1902.[51] Organizations such as the German League for Bird Protection mobilized a huge constituency; other groups focused on particular areas: the Isartalverein [Isar Valley Society] is one such group which Dominick traces from its foundation in 1902 to its present-day existence.[52]

By the end of the nineteenth century, however, nature conservation had become an overtly political issue. There is a distinction to be made between conservative, *völkisch* nationalists[53] operating in the tradition of Riehl on the one hand, and socialist commentators and conservation organizations on the other. As we have seen, the link between conservation and nationalism identified by Arndt was later amplified and popularized by Riehl. Three of the most important figures to perpetuate and develop the views set out by Wilhelm Heinrich Riehl in mid-century were Ernst Rudorff, Hugo Conwentz and Paul Schultze-Naumburg. In 1880 Rudorff published an essay entitled 'Über das Verhältnis des modernen Lebens zur Natur' ['On the relationship of modern life to nature'] in which he argued that the source of German culture lay in the forests and landscape of the homeland: 'The very roots of the Germanic character lie in an intimate and profound feeling for nature.'[54] The essence of Germanness was under attack not only from industrialization, but also from agricultural rationalization, he claimed. Towards the end of the nineteenth century Rudorff's views found broad resonances. He combined mystical nationalism with a reverence for tradition, and for an idealized past to which industry – and urban growth – were anathema. This combination of romantic conservatism and anti-progress rhetoric brought its own problems. Reluctant to criticize the industrialists who were delivering Germany's powerful position in the world, Rudorff's demands for the protection of the environment – which for him included the built environment, provided it fitted into an idealized notion of tradition – were limited in scope. He favoured the creation of protected areas which would remain immune from the changes taking place.[55]

This reluctance to challenge the primacy of progress is, if anything, even more pronounced in Hugo Conwentz, an almost evangelical advocate of the importance of protecting nature who headed the Prussian state conservation agency. In 1904 he produced a report for the Prussian government on *Die Gefährdung der Naturdenkmäler und Vorschläge zu ihrer Erhaltung* [*The Endangerment of Natural Monuments and Proposals for their Preservation*] which alerted the public, and state governments, to, as Dominick puts it, 'the manifold dimensions and causes of damage to many aspects of the environment'.[56] Conwentz proselytized on the subject with such assiduity that by 1914 he had succeeded in 'mobilizing a sizeable and durable constituency for conservation'.[57] As in the case of Rudorff, however, Conwentz's plea was for preservation of endangered areas or features, rather than for industry to make sacrifices in favour of the environment. Even in the above-mentioned report, he was careful to enter a caveat: 'There should be no question of pushing back industry a single step in order to preserve notable scientific and aesthetic features of nature.'[58] He goes on to say that industry should be capable of finding ways of mitigating its effects on nature. Like Rudorff, Conwentz was reluctant to put real pressure on industry to reform its practices. This was not from want of nationalist support in the cause of nature conservation.

By the beginning of the new century, Rudorff and Conwentz had become founder members of the Bund Heimatschutz [League for the Protection of the Homeland], whose first chairman was Paul Schultze-Naumburg.[59] Founded in 1904, the Bund Heimatschutz was an umbrella organization which embraced 250 affiliated groups by 1916. Nature conservation was only one of the League's preservationist aims, others including traditional architecture, folk customs and art.[60] The conventional wisdom has been that the League's concerns were hardly environmental in the modern sense, but mainly conservationist, in keeping with the views expressed by two of its founders as outlined above. Andersen has argued that, in keeping with its reactionary origins and its Romantic idealization of nature, the League judged the preservationist value of natural features mainly on aesthetic grounds. Its opposition to the building of a hydroelectric power station at Laufenburg on the Rhine, for example, took the form of a proposal to resite the project somewhere less aesthetically damaging. Failing in its opposition, the League was reduced to hoping that the design of the power station would be as visually pleasing as possible.[61] Andersen's view of the Heimatschützer as an ineffectual campaigning group is contested by

Rollins, who suggests that their 'aesthetic communication contributed directly to popular environmental consciousness by proclaiming the value of beautiful and healthy landscape decades before science could prove such value.'[62] While agreeing that the aesthetic aims of the movement were paramount, Rollins argues that League policies on the importance of hedging and mixed forest, and their understanding of the dangers of pollution, for example, do indeed indicate an understanding of environmental problems and a determination to take action to solve them which is recognizable in terms of modern environmentalism. Rollins's well-researched efforts to rescue the League from condemnation do result in a more differentiated view of the organization. But we are left with a fundamental problem: Rollins argues that the League was forced to rely on the aesthetic arguments in the absence of scientific evidence.[63] Yet it is precisely the advent of convincing scientific evidence which allowed the environmental movement to blossom in the form in which we know it today. There seems little point, then, in trying to show that the League was in fact environmentally minded in any modern sense. At this stage we are still confronted with a concern for the preservation of natural features, albeit a concern which was, as Rollins shows, remarkably well-informed. Rollins's most convincing argument in favour of the Bund Heimatschutz as pioneering environmental campaigners is the way in which it raised the level of interest in damage to the natural environment. But in this respect it was by no means alone.

Conservatism and *völkisch* nationalism were not the only ideological impetus for nature protection in Germany at this period. Engels's description in *The Condition of the Working Class in England* of the filthy state of the Irk, and of the working-class districts of Manchester in general, graphically identifies the environmental consequences of unchecked capitalism. But the victims are human, of course, and the cause is not industrialism *per se*. Blühdorn asserts that Engels's *Dialectics of Nature* defined 'what today would be called the environmentalist approach';[64] but it had little contemporary influence. For Marx, capitalism dehumanized its victims, removing them from the the cycles of nature and its harmony.[65] Anna Bramwell has examined in detail the question of whether Marx can be claimed as an early ecologist, concluding that both industrialism and anthropocentrism are so central to his thought that ecologism (or even environmentalism) are excluded.[66]

Jost Hermand is more sympathetic, arguing that Marx's recognition of the limited resources of land as an economic factor led him to avoid making prognoses for the classless society he propounded.[67]

A socialist politician with no such reservations, however, was August Bebel (a close friend of Engels). Both Jost Hermand and Raymond Dominick have drawn attention to the environmental content of Bebel's *Die Frau und der Sozialismus* [*Woman and Socialism*], which was first published in 1879 and ran to over fifty editions with translations into English, French and many other languages. This enormously popular book argued passionately in favour of the advantages of socialism for society in general and women in particular.[68] It also contains passages on deforestation, soil erosion and flooding, and on the dangers to workers' health of pollution. As one would expect, Bebel is quite clear that private enterprise, and its refusal to incur the necessary costs, is at fault in refusing to solve the problem of 'dust, smoke, soot and smells'.[69] It seems, then, that in Bebel's vision, the socialist society would be a cleaner one. More remarkably, Bebel considers the question of finite fossil resources, raising the possibility of renewable energy sources and citing 'our waterways, the tides, the wind and sunlight'.[70] He even holds out the prospect that in the future geothermal energy might be a possible alternative source of energy. These ideas were certainly progressive (in fact many still are), and at first sight one appears to be confronted with an extraordinary prophet of green revolution. But there is nevertheless a need for caution in terming Bebel's ideas ecological, or even necessarily environmentalist in a modern sense.[71]

For example, Bebel's desire for well-ventilated, unpolluted factories was a natural and commonplace part of the socialist prescription for change; the passage appears in a section promising the alleviation of bad working conditions under socialism, rather than on environmental depredations. His concerns should certainly not be dismissed merely because we would today view them as (understandably) anthropocentric. At the same time, however, Bebel argues strongly in favour of increasing both production and productivity;[72] renewable energy sources are seen as a way of removing the limits to growth imposed by coal reserves, rather than of restricting economic growth.[73] The alternative energy sources he cites include hydroelectricity; ironically, hydroelectric projects probably drew more protests from German environmental

campaigners between 1900 and 1960 than almost any other industrial proposal.[74] Moreover, Bebel was not alone in looking for
alternatives to fossil resources; the problem was widely discussed in
Germany during the 1880s.[75] Bebel was greatly enthusiastic about
the prospects of technological and industrial progress, and about
electricity in particular: he quotes scientists who see the possibilities
of making food from stone and wood using this 'most powerful of
all natural forces'.[76] The ideas on solar and geothermal energy were
drawn from a lecture delivered in 1894 by the French chemist
Marcelin Berthelot on the future technological prospects of the
chemical industry.[77] By the year 2000, Bebel reports, agriculture
will no longer be necessary, since all food will be made through
chemical processes: 'What plants previously did, industry will do,
and it will do so better than nature.'[78] People would carry a can of
chemical food containing all necessary nutrients. Yet this most un-
green prospect gives rise to a utopian vision where the world,
devoid of agriculture, will become a 'garden, in which grass and
flowers, bushes and forest can be allowed to grow as desired, and
in which humankind will live amid abundance, in a golden age'.[79]
Bebel leaves readers free to draw their own conclusions about these
speculations, but there can be no doubting his own faith in the liberating potential of industrial progress. This is an excellent example
of the difficulties one encounters in attempting to perceive proto-
green ideas in terms of our modern environmental understanding.
There are unquestionably green elements to Bebel's vision of the
socialist future, but the green utopia is to be brought about and
sustained by methods which appear decidedly non-environmentalist.

Perceptive and widely disseminated though Bebel's views were,
there is no evidence that this part of his political creed had any
practical effect, or that it was seized on by proto-ecological
campaigners. There were, however, environmental campaigners
proper on the left of the political spectrum who, while smaller in
numbers, were arguably more effective both environmentally and
politically than their conservative counterparts. Affiliated to the
socialist SPD, the Naturfreunde [Friends of Nature] were formed as
a tourist club in Vienna in 1895.[80] The club was composed of
walking enthusiasts and so-called 'Walzbrüder', itinerant skilled
workers following the old tradition of walking to where the work
was. These workers were able to disseminate political propaganda
as well as their skills. In contrast to the Heimatschützer, the

Naturfreunde were, of course, unafraid of criticizing capitalism, and especially private ownership. In 1906 they began a campaign for free access to a countryside, large tracts of which lay in private hands. Ironically, in stressing the need for common ownership of the natural environment, the Naturfreunde were partially echoing the prescription of Wilhelm Heinrich Riehl (see above), who had so inspired the Heimatschützer. Nevertheless, their journal *Der Naturfreund* [*The Nature-Lover*] urged the association's members to take part in an escalating series of protests ranging from legitimate lobbying of parliament to civil disobedience in the form of mass trespass. Whether the desire to allow large masses of people access to otherwise sparsely populated countryside can be termed environmentalist is debatable, whatever one may think of the demerits of private property. But it is clear that in resorting to extra-legal methods of protest, the Naturfreunde were foreshadowing modern environmental groups in ways unmatched by the conservative wings of the movement. Their protests were directed at a whole range of environmentally damaging projects from deforestation and industrial development to quarrying and peat-moor plundering. The Naturfreunde were also ahead of their time in making an explicit connection between capitalism and the exploitation of natural resources. 'It's incredible how brazenly capitalism tries to get its claws into everything', thundered *Der Naturfreund* in 1912, when it was revealed that a new quarry was to be dug and a river dredged in order to transport the resulting gravel.[81] On occasion the Naturfreunde did make common cause with the Heimatschützer, and did face similar problems. Just as industrial interests were prone to treat the League for the Protection of the Homeland with a certain contempt, so were the Naturfreunde regarded with suspicion by their Social Democratic mother party, which itself relied on industry – only the ownership and management of which was in dispute – to deliver the aspirations of its members. There were groups, however, which sought to reject industrial society altogether.

As the new century approached, numerous *Lebensreform* [literally, 'life-reform'] groups came into being as a reaction against the increasing urbanization and technologically alienating nature of *fin de siècle* Europe. This was by no means a uniquely German development, and indeed by its very nature the radical life-reform and commune movement was the least bound to a specifically German tradition of those so far discussed. In Britain, 'simple life' and 'back-to-nature' movements were given impetus by the visions of John Ruskin and William Morris, whose arcadian utopia *News from Nowhere* was translated into German by Wilhelm Liebknecht in 1892.[82] One strand of the life-reform movement developed from the beginnings of homoeopathy. Urbanization and industrialization were held to exert deleterious effects on health, which could be ameliorated by resorting to natural methods. In the main, this strand consisted of groups which practised abstinence from harmful man-made substances or artefacts: alcohol, smoking, even spices were abjured.[83] Vegetarianism, though not new, gained in popularity, while nudism represented an effort to regain the innocence of the natural state. Rudolf Steiner's creed of anthroposophy was perhaps the most influential and enduring.[84] Adherents of such creeds, rejecting the breakneck industrialization which propelled Germany to the top of the economic growth league by 1914, formed communes to practise their beliefs. However, one can hardly talk of a movement in a coherent sense, so atomistic was the phenomenon and so short-lived the individual communes. Hundreds of groups existed in Germany practising the natural way of life around the beginning of the century.[85] Indeed, colonies practising the simple life were set up all over the world. Nietzsche's brother-in-law famously founded a colony in South America, whose descendants still live there.[86]

The groups associated with the Swiss village of Ascona are amongst the best-remembered, perhaps because of the fame of some of those involved.[87] Ascona was founded by the brothers Gusto and Karl Gräser in 1900.[88] Their initial colony lasted only a year, but started a tradition, and led to the establishment of the sanatorium Monte Verità, which prescribed 'vegetarian food, sun baths, air baths, earth cures, and water cures' for its patients, who included Hermann Hesse. The treatment was more than homoeopathic: nature cures went together with cult celebration of air, water and light.[89] Asconans were not environmental campaigners. Nature worship was, if not devoid of political content, then at least not overtly political, and *Naturmenschen* [nature people], in rejecting society, simultaneously had to renounce much potential influence. In any case, their long hair, bare legs and sandals more often than not made them objects of derision.

One might, then, reasonably ask what significance such groupings have in a historical survey of green ideas.[90] Their influence was slight and their members marginalized. But their importance lies in the fact that their ideas contain important elements of ecologism, and that their chosen method of expressing their ideas lay in practising them on a small scale. The very atomism and decentralization of these groups made them into a kind of prototype for later views of how an ecological society might look. The case must not be overstated: many of the adherents of these ideas were not only mystical, but thoroughly anti-science (the Gräsers, for example). Modern ecologism is inconceivable without a scientific basis to understand the world. Further, they lacked the kind of political manifesto without which radical ecologism has no blueprint for change.

The publication in 1913 of an essay entitled 'Mensch und Erde' ['Man and earth'] by Ludwig Klages provided the communalists with an impassioned case in favour of their attempts to rehearse an alternative way of living on the earth. The vitalist philosopher Klages rejected technology vehemently. In 'Mensch und Erde', Klages not only excoriates the destructive effects of industry on the environment, but apocalyptically accuses industrialists of matricide; they will end by destroying mother earth. Industrial technology is equated with the destruction of life. Man had 'fallen out with the planet that bore him and nourishes him'.[91] In 1913, Klages recognized and articulated the possibility that mankind was exploiting the *global* environment in an unsustainable way.

After the First World War, alternative life-style groups began to take on a more political and distinctively ecological tinge, a development for which the anarchist Gustav Landauer was in some measure responsible. There is no need here to go into detail on Landauer (see chapter 3 of this volume). Suffice it to say that his adaptation of the ideas of Kropotkin and Tolstoy on socialist communes exerted an influence not diminished by his reluctance to become personally involved in the settlement movement. His 'Aufruf zum Sozialismus' ['Call to socialism'] (1911) proposed a form of non-Marxian, agrarian socialism which found an echo in the attempts of anarcho-syndicalist communalists to initiate a rural green socialist revolution.[92] In January 1921 settlement revolutionaries met at Heinrich Vogeler's Barkenhoff commune at Worpswede to plan a massive demonstration. Its aim was to seize control of the means of food production and to bring about a change in the way people lived their lives: 'bread, sun, light' was the slogan. The two main initiators of the action were Leberecht Migge and Paul Robien. Migge was already practising settlement in the form of his Sonnenhof colony. Ulrich Linse has made a convincing case that Robien should be regarded as 'the first radical German "Green"'.[93] The crucial feature which distinguishes Robien, Migge and other anarcho-communalist greens from green socialists such as Bebel was their rejection of Marxism for a decisively anti-industrial stance. Robien's renunciation of human interference in nature was indeed radical. While Migge was willing to countenance thorough-going cultivation of the countryside, Robien was determined that cultivation should be kept to the minimum necessary, and that as much countryside as possible should be left in a wild state.[94] Robien's vision of the liberation of workers seems a seductive one:

> They are emerging from the soot of an industrial hell. They want to escape this diabolical embrace, get out into the open air, eat their own bread, pure bread, and work in sun and light . . . Anyone who has seen the bitterness and determination in the eyes of these comrades will understand that all is not lost with the masses.[95]

The planned action, however, was a miserable failure, the group's influence on their contemporaries slight. But there are enough reasons to see precursors of radical political ecologism in Germany in this group. Migge and Robien articulated many of the standard elements of ecologist economics that we know today. Migge's 'Das grüne Manifest' ['The green manifesto'], published in 1919, recommends a series of measures to promote the productivity of land use in a sustainable way, drawing on a whole range of organic cultivation methods. The aim was to reach a society which can live off the land using recycling to renew resources, stretching small technology to its limits to do so. His Sonnenhof commune put these ideas into practice.[96] Robien remained a tireless campaigner for ecosocialism, stressing, in contrast to the nationalist Heimatschützer, that an internationalist approach was essential, and maintaining that the green vision absolutely necessitated a reduction in consumption. He also argued, on a scientific basis, that animals and human beings had an equal right to existence.[97] That said, Robien had a side which would be less welcome to modern Greens. He despised Jewish socialist leaders (such as Bebel) for promoting the 'city revolution', and put much of the blame for the ravages of capitalism on Jews. His crude anti-Semitic remarks included the chillingly prophetic suggestion that '[Jews] should be treated like game and done away with'.[98]

Robien's anti-Semitism did not go unanswered among fellow anarchists: he was condemned as a racist by leading contemporaries. There does, however, seem to be a fatal connection between green ideas and racial ideology during this period. One need not ascribe this to some quality inherent in green ideas: racial ideology had been penetrating many aspects of European culture since before the advent of any coherent conservation movement. Like any fashionable theory, it was adopted by adherents of all kinds of disciplines to explain their particular interest. And despite Robien, racism was much more associated with *völkisch* nationalism than with any other strand of proto-environmentalism: conservationists had been calling for the preservation of German racial purity in the homeland as early as 1913. Others compared the impact of air pollution on forests to that of immigrants on the German population.[99] This is not to say that all Heimatschützer were necessarily racists.[100] But there is enough evidence that prominent conservationists eagerly embraced ideas which were being popularized by the National Socialists (NSDAP) during the 1920s.

Paul Schultze-Naumburg is an excellent example of the way in which Heimatschützer adopted racial ideology. Bewildered by the conservation campaign's failure to prevent the incremental decline of the natural and (as he saw it) of the built environment, he began during the course of the 1920s to ascribe this failure of popular support to the genetic make-up of the *Volk*, rather than to apathy or differing priorities. By 1930 at the latest, Schultze-Naumburg had come to the conclusion that social equality and medical advances were interfering with natural selection: 'The profound causes of the phenomenon, observed on all sides, that our whole environment is becoming ever more cheerless and ugly . . . cannot be explained other than through an increase in lower-quality stock via procreation.'[101] The solution, Schultze-Naumburg proposed, was not only eugenics, but euthanasia. Sieferle has pointed out that this argument had the advantage of relieving industry of responsibility for environmental degradation and depositing it with the unwashed masses.[102] At the end of the 1920s, Schultze-Naumburg threw in his lot with the NSDAP, joining a group in Munich which included Alfred Rosenberg and Richard Walther Darré. His country house became a favoured venue for party activists. In 1930 he joined the party and became a founder member of Rosenberg's Kampfbund für deutsche Kultur [Militant League for German Culture], which paved the way for the Nazi attempt to appropriate German culture. In 1932 Schultze-Naumburg, president of the Bund Heimatschutz from 1904 to 1914, was elected to the Reichstag as a National Socialist delegate.[103] The convergence between Heimatschutz and Nazism was complete: the League was *gleichgeschaltet* [absorbed into the Nazi party organization] in 1933.

Nature conservation was not the only area in which National Socialism was able to exploit green ideas. The settlement movement fed into the policy of 'Ostkolonisation' which was popularized by Hans Grimm in his novel *Volk ohne Raum* [*People without Living Space*] of 1926, and which was put into practice in Poland during the early 1940s. The Nazi movement drew on *Lebensreform* too; theories about clean, healthy living would help the Germans to become the fittest and best-adapted of the Nordic races. But it was the ideology of 'Blut und Boden' [Blood and Soil], formulated in the main by Walther Darré, which adapted, even perverted, elements of the early green tradition in Germany to form a National Socialist political weapon much used in the early years of the regime.

From its beginnings, the NSDAP found natural allies in the countryside lobby, and, indeed, in the farmers' lobby. The *Bauern* were a powerful constituency, disgruntled with the taxation policies of successive Weimar governments, whose support was important to the movement's success. Although Darré had not belonged to Hitler's core group of long-standing activists, he quickly found favour with his elevation of the German *Bauer* [farmer or peasant] to the new Nordic aristocracy. Furthermore, he had struck up a friendship with Himmler, when the future head of the SS liaised with the Artamanen, a right-wing settlement movement which similarly subscribed to back-to-the-land ideas. Darré postulated a mystical relationship between German peasants and the soil they tilled stretching back generations. The *Bauern* were the determining factor in realizing the German *Volk*'s historic destiny; they embodied and protected the essence of Germanness. Darré published two books popularizing these views: *Das Bauerntum als Lebensquell der nordischen Rasse* [*The Peasantry as Life-Source of the Nordic Race*] (1929) and *Neuadel aus Blut und Boden* [*A New Aristocracy from Blood and Soil*] (1930). After 1933 his works ran to many editions. Darré only joined the party in 1930, having been offered the job of running agricultural policy. From 1931 he was made head of the SS Rasse- und Siedlungshauptamt [Central Office for Race and Settlement]. Darré was thus able to combine his crude racism (derived in part from his study of animal breeding) and anti-Semitism with a powerful belief in the importance of organic farming acquired during his study of agriculture in the 1920s.

In 1933 he was made *Reichsbauernführer* [peasant leader] and Reich Minister for Food and Agriculture; a host of other posts and honours followed. For the first time, then, organic methods of agriculture had the imprimatur of a state minister. More than this, Darré saw an opportunity to put into practice the racist, mystical, anti-urban, anti-industrial, indeed anti-capitalist views he had been propagating for years. His bizarre theories on the genetic destiny of German farmers were enshrined in the *Reichserbhofgesetz* [Reich Estate Inheritance Law] of 29 September 1933, which debarred anyone with 'Jewish or coloured blood' from inheriting farms.[104] He persuaded Hitler and Himmler that his back-to-the-land policies would make Germany self-sufficient in food. Although organic farming was indeed promoted for some years, Darré's attempt to reverse the long-standing trend of movement from the countryside to the cities failed. Indeed, the trend towards urbanization accelerated. The gap between actual state policy and professed beliefs was as great in this part of the National Socialist state as elsewhere. Darré could not resolve the irreconcilable contradictions which the power imperatives of Hitler's policies demanded. *Blut und Boden* required not only an intimate bond with nature, but also more *Lebensraum* within which to practise that bond.[105] The land could only be acquired by war, and wars could only be fought with the support of capitalist financiers and renewed heavy industrial output. Even before the beginning of the war, the result was an onslaught on the natural environment in the form of autobahn-building, military construction, land-draining, further industrialization, urbanization and so on.[106] Although attempts were made to improve the aesthetic impact of autobahn construction, and to manage ground-water resources properly, meaningful environmental considerations came a long way behind military and economic priorities. *Blut und Boden*, with its powerful emotional appeal, was for Hitler little more than a useful but short-term political tool. Even organic farming was eventually abandoned as higher agricultural yields were demanded and Darré's position weakened with the onset of war. He lost his post in 1942.[107]

Nature conservationists saw the NS regime as an opportunity to realize long-held dreams, even if they did not wholeheartedly subscribe to the ideals of the new Germany.[108] Werner Sombart, for example, rejected racial ideology, but perceived the socialist element of National Socialism as an opportunity to pass stringent legislation limiting the environmental excesses of capitalism.[109] At the end of the 1930s he was disillusioned. By contrast, Walther Schoenichen, who in 1954 was to write the first history of German environmentalism, was an enthusiastic adherent of racist, *völkisch* nationalism throughout the 1920s. As director of the Prussian conservation agency he took part in the *Gleichschaltung* of existing conservation groups in 1933 under the Reichsbund für Volkstum und Heimat [Reich League for Volkstum and the Homeland]. His book *Naturschutz im dritten Reich* (1934) held out great hope for a green future in the new Germany. Although Schoenichen's vision went the way of Darré's, certain measures in the area of nature conservation did pass into legislation during this period. Many were repressive, of course; the socialist Naturfreunde, for example, were forcibly dissolved in 1933. Measures to prevent cruelty to animals, passed in 1933, were tailored to suit the pre-eminent anti-Jewish campaign: Jewish ritual slaughter was outlawed while scientific testing on animals was permitted. Other legislation, however, seemed to give the Heimatschützer what they had been campaigning for since the beginning of the century. Measures were taken between 1933 and 1934 to protect wild animals, to ban advertising hoardings from the countryside, and to establish new parks and nature reserves. The main legislation was passed in 1935 in the form of the *Reichsnaturschutzgesetz* [Reich Nature Protection Law], which provided sweeping national powers to co-ordinate previously autonomous local environmental agencies and, in effect, to protect any aspect of the natural environment which might be deemed in need of such protection. While the legislation was fulsomely received by traditional conservationists, its effects were comparatively limited, given the real priorities of the Hitler state. Nevertheless, the law remained on the statute book until 1958, when its powers were devolved to the individual German *Länder*.

Despite the catastrophic human and environmental cost of National Socialism, the defeat of Germany did not significantly interrupt conservationist activity. Regional and special-interest nature-conservation groups continued their work, within constraints imposed by circumstance and the occupying powers, and regional governments continued to consider the needs of the natural environment. The most senior government nature-conservation officer, Hans Klose, retained his post from 1938 through into the post-war period.[110] Although, as Dominick has shown, in 1945 membership of leading conservation groups was approximately half what it had been in 1940, the groups remained in existence and by 1965 had almost recovered their previous levels.[111] Similarly, protests against industrial projects continued to take place, particularly against hydroelectric power stations (reprising the pre-First World War origins of the Heimatschutz movement), from 1947 through to the late 1960s. Clearly, conservationists faced an uphill battle in a country engaged in post-war reconstruction and showing record growth figures during the 1950s. Yet the rapid spate of industrial growth, coupled with a press determined to assert its freedom in the new democracy, raised levels of environmental awareness to new heights: incidences of air, water and noise pollution made lurid headlines in the 1950s. But the decisive innovations, which were to change the face of environmentalism and give rise to genuine ecologism were, on the one hand, advances in scientific understanding of the nature and consequences of ecological disruption, and on the other, the nuclear threat.

By the mid-1950s, the threat of nuclear annihilation was genuine and apparent to all. If the Cold War became heated, Germany would be the battlefield: for the first time in human history, the apocalypse was at hand in a practical sense. The rearmament of Germany in 1956 coincided with the decision to build nuclear reactors for domestic use. Initially, conservationists welcomed atomic power as a way of avoiding any need for future hydroelectric power stations. Curiously, it was the electricity industry that was suspicious of the new technology: its reliability could not be guaranteed.[112] By the late 1950s, however, protests against atomic energy as well as the bomb were taking place in Germany as elsewhere. The activist Bodo Manstein publicized the danger of radioactive emissions in detail, drawing attention particularly to the possibility of meltdown,[113] and demonstrations against nuclear reactors took place all over Germany.[114]

But nuclear power was not the only issue. Awareness of environmental dangers of all kinds, and of their consequences for global ecology, had been growing throughout the 1950s. Writers such as Erich Hornsmann, Erwin Gamber, Reinhard Demoll, Ernst Hass and Peter Härlin warned of the dangers not only of deforestation, monoculture and erosion, but also of pesticides and other chemical pollution, of nuclear energy, of radioactive fall-out in the event of nuclear war, of traffic, population pressure, and of diet.[115] Moreover, in keeping with the apocalyptic approach which was beginning to supplement the more narrow conservationist concerns traditionally present in Germany, the ultimate result was to be the death of humankind. It was during this period, then, that some of the main elements of modern ecologism were coming together in Germany: a detailed, scientific understanding of the global effects of human impact on the environment, and the prediction that the end result would be the annihilation of humankind. What was missing was a detailed blueprint for change. That was supplied by Günther Schwab in 1958.

In *Der Tanz mit dem Teufel* [*Dance with the Devil*] the global ecological crisis is dramatized (rather archly, it must be said) in the form of an interview between the devil, his helpers and four human characters. The premise is that human beings are doing the devil's work at his instigation by destroying the environment to the point where life (not just human life) is untenable. Allusions to the death of God, the supremacy of Satan and the presence or absence of angels are common in the Germany of this period, where the temptation was to seek explanations for the catastrophe of Hitlerism outside the sphere of human activity. In the case of Schwab, however, it is clear that human beings are very much to blame, and enormous scientific detail is provided to support an argument which is more or less the opposite of anthropocentric. Schwab supplies a voluminous bibliography to support a thoroughgoing attack on air pollution, water resource depletion and pollution, noise pollution, dietary deficiencies and poison in foods, the destruction of forest, agricultural land and traditional agricultural practices, the potential disaster of nuclear proliferation, of population pressures in the face of limited resources, of global warming, and of other dangers too numerous to mention. In many cases figures and sourced quotations are provided. Moreover, the dramatization ends with the destruction of the earth, survived only by one man and one woman, who resolve that the new Eden will rely only on ancient, natural tradition. Leaving the artful sentimentality aside, this work is fascinating for a number of reasons.

This is a work written for a popular audience which, as early as 1958, would allow anybody to gather the necessary knowledge to be fully aware of the global environmental crisis, and of the apocalyptic consequences. The literal use of devil's advocacy allowed the soothing arguments of authorities anxious to defuse fears (especially about the nuclear danger) to be set against the contemporary state of scientific knowledge. Furthermore, Schwab has no compunction in laying the blame for the problems at the door of governments and multinational corporations, foreshadowing later analyses. And the book was widely received: it was reprinted in a fifteenth edition, in unchanged form, in 1991. Perhaps most importantly, all these ingredients add up to a genuinely ecologist view: it is holistic, scientifically based yet critical of science, apocalyptic and prescriptive. The radical change it proposes to avert catastrophe, however, reveals a darker side to Schwab's work. It is impossible on occasion to overlook echoes of Richard Walther Darré, who had died only five years earlier in 1953.

Salvation from the environmental crisis is here to be found in the 'ewiges Bauerntum': 'The higher powers, the lasting and creative powers, are rooted in the eternal peasantry.'[116] A government minister who alludes to the obsolescence of *Blut und Boden* mythology is mocked. There is little evidence of German nationalism in this global analysis (Schwab was in any case an Austrian who originated in Czechoslovakia), but the blueprint for change has recourse to the '100,000-year-old soul of the peasant' in a barely veiled reference to Darré's theories. Though simplistic, the recommendation that the economy and society of the future should be given over to smallholdings and subsistence farming is in tune with elements of radical ecologism. However, the mystical celebration of the peasantry is backward-looking and paradoxical, given the modernity of much of the analysis. The ideological slant of *Der Tanz mit dem Teufel* emerges in other ways: the crude condemnation of the United States as a degenerate society, and the anecdotal nature of much of the anti-American rhetoric, for example. There are sections in the book which border on the fascistic: the chapter on population more or less recommends the death of billions by ruthless natural selection: 'Today there are two and a half billion people too many in the world, people who have no right to life because they managed to acquire life by stealth and against the laws of nature.'[117] Attacks on egalitarianism are powerfully reminiscent of Schultze-Naumburg at his most vehement, and the argument culminates in a series of blatantly racist assertions. The introduction of modern medicine and hygiene had allowed the population of India and other countries to grow limitlessly; 'der gute Tod' [good death] had been prevented from exercising its beneficial effects: 'where there is no selection, life itself falls ill.'[118] The result of all this is that an ever-smaller perceptive élite is confronted by a swelling, uncontrollable mass of the uneducated and uncultured. Europe would eventually be overrun by 'mongols', who would be succeeded by 'yellows'; America would succumb to 'browns'.[119] At the end of the book, the Adam-and-Eve survivors are Europeans. The ideological slant of this early example of ecologism, then, is rooted in a wholly unwelcome tradition.

This does seem to suggest that the link between far-right ideology and green ideas – in this case practically indistinguishable from post-1972 ecologism, except in political ideology – did not end with the Nazis. The political significance of that insight is a moot point: there is little evidence that Cassandra-like warnings of this kind had much practical effect.[120] For example, a green manifesto of remarkably advanced character with impeccable credentials, based on the Federal Republic's Basic Law, sank more or less without trace after its publication in 1960.[121] Nevertheless, it is possible to trace a steady interest in environmental awareness in Germany throughout the 1960s, probably impelled by the publication of Rachel Carson's *Silent Spring* in 1962 and its translation into German the following year.

Furthermore, the post-war period saw a new approach to local environmental crises, that of direct action at the grass roots. As early as the mid-1950s, citizens' initiatives were launched against incidences of road-building, air pollution and water pollution.[122] Through analysis of press coverage and membership of conservation and environmental organizations, Dominick has shown that 'national concern with environmental problems experienced a major spurt from the mid-1950s to the mid-1960s, followed by a lull of three or four years.'[123] The post-war interest in the environment, then, usually dated to about 1970, was in fact a renewed upsurge after a previous peak of interest about 1959. When, in 1972, the publication of *The Limits to Growth* transformed the terms of green debate, there was in Germany already a substantial constituency of support, a relatively high level of environmental awareness, a strong tradition of green ideas and a long, continuous tradition of proto-green activism. The history of the development of green ideas, awareness and politics after 1970 has been extensively documented; there is no need here to augment an already voluminous literature.[124] The question which I want to pose is this: which, if any, of the traditions outlined in this chapter did the German Green movement in general, and Die Grünen in particular, draw on after their foundation in 1979?

Where green activists have drawn directly on historical documents, the influences are easy enough to see. For example, Ludwig Klages's 'Mensch und Erde' of 1913 was reprinted in 1980 with a foreword by Bernhard Grzimek, a prominent post-war pioneer of the German neo-green movement.[125] But one searches most modern green creeds or political manifestos in vain for the names of early environmental commentators or activists such as those that appear in this chapter.[126] And merely establishing that there *were* historical precedents for modern green protest and activism does not automatically make those precedents political ancestors. It is difficult to see any clear connection between, say, Arndt, Humboldt or Riehl drawing attention to deforestation in the first half of the nineteenth century, and green politicians protesting against the devastation of Amazonian jungle in the 1980s. No doubt many were unaware that such predecessors existed; is it then meaningful to say that they stand in a tradition? And since the history of environmentalism is largely one of defeat, failure and marginalization, one could hardly argue that green activists might either boost their morale or learn techniques of success by a study of proto-green movements. After all, if the warnings of earlier generations had been heeded, there would have been no need for a modern green movement.

Nevertheless, there are still questions to be asked. The thorniest problem is the extent to which the nationalist, *völkisch* tradition, and in particular, the National Socialist variety, might reasonably be regarded as a precursor of green ideas in modern Germany. Anna Bramwell has probably been the most forthright in this respect. An article she published in 1984 in *History Today* bears the title 'Was this man "Father of the Greens"?', above a picture of Darré in Nazi regalia.[127] The answer to that question (though not given in the article) must, of course, be no. Clearly, no single figure could be regarded as the progenitor of modern Greens in Germany or elsewhere. As we have seen, some aspects of ecologism have an ancient lineage, and even environmentalism has a long history of its own. Darré himself stood in a tradition. The broader question might be: how important is the green tradition within *völkisch* nationalism for an understanding of modern green politics? It is true to say that Heimatschützer and Naturschützer mobilized a large constituency throughout the twentieth century. That is to say, there was by 1970 a large constituency in Germany sympathetic to green ideas in the broadest sense, and a substantial proportion of that constituency was rooted in a conservative tradition with *völkisch* antecedents. But there are no grounds for suspecting some direct link to Die Grünen. None of the authors of the main tenets of green political theory and practice who remained in the party after 1980 would have the remotest sympathy with such a tradition. And the Greens are not the only representatives of broadly ecological/environmentalist politics in Germany; the variety of groups and emphases remains as bewildering today as a century ago. So there seems little reason to try to argue that Die Grünen are the inheritors of a right-wing tradition when there are right-wing green parties, and even eco-fascist groupings, in the Federal Republic today, who can be placed in precisely this tradition without difficulty.[128]

The modern German Green Party's most immediate ancestor is the student movement of 1968; its pedigree is radically left-wing. It thus makes more sense to look for precedents in the ecosocialist tradition; and more especially, given their radical nature, among the eco-anarchists. Yet even here it is not helpful to speak of political ancestors. The green movement in contemporary Germany is a new phenomenon, which itself has undergone radical mutations since 1970, and continues to do so. Ingolfur Blühdorn argues that changing concepts of nature have brought about a sea-change in environmental politics in the 1990s, when political ecology is a concept shared by descendants of groups originally from very different political traditions.[129]

Equally, it would be wrong to say that National Socialism has had no effect on the German Green Party. In 1980 constituting delegates famously resolved on four pillars of principle for the party programme, only one of which refers to the environment. The

Greens aimed to be ecological, grass-roots democratic, socially responsible and non-violent. Clearly, the last three terms specifically make impossible any affinity with fascistic politics. They are a patent reversal of Nazi values: genuine democracy, not authoritarianism; care of the weak, not survival of the fittest; peace, not war. The most recent programme (writing in 1997) of Bündnis 90/DIE GRÜNEN adds human rights and gender equality to those pillars. The German Greens, then, are obviously diametrically opposed to Nazism and fascism. Does this mean that they are liberated from the legacy of Nazism? No: responsible stewardship necessitates historical awareness. It would be as wrong for green activists to deny the historical links between green ideas and fascism as it would be to argue that green ideas are necessarily fascistic, or stand exclusively in a left-wing tradition. If green politics is about anything it is about acknowledging responsibility: to the past as well as the future.

Notes

[1.] 'A historical perspective needs to be defined for the environmental debate' is the argument in *Besiegte Natur. Geschichte der Umwelt im 19. und 20. Jahrhundert* [*Conquered Nature: A History of the Environment in the Nineteenth and Twentieth Centuries*], ed. Franz-Josef Brüggemeier and Thomas Rommelspacher (Munich, 1989), 6. (This and all other translations throughout are mine unless otherwise indicated.) A similar point is made in David Pepper, *The Roots of Modern Environmentalism* (London, 1989; first published 1986), ix, and in the back-cover blurb to Jost Hermand, *Grüne Utopien in Deutschland. Zur Geschichte des ökologischen Bewußtseins* [*Green Utopias in Germany: On the History of Ecological Consciousness*] (Frankfurt am Main, 1991). Also: 'For most of us, even the committed activist, the green movement has no history' (Derek Wall, *Green History: A Reader in Environmental Literature, Philosophy and Politics* (London and New York, 1994), 1). In fact, a history of conservation movements in Germany appeared as early as 1954. See Walter Schoenichen, *Naturschutz, Heimatschutz. Ihre Begründung durch Ernst Rudorff, Hugo Conwentz und ihre Vorläufer* [*Protecting Nature and the Homeland*] (Stuttgart, 1954).

[2.] This is the title given to the first chapter in Ulrich Linse, *Ökopax und Anarchie. Eine Geschichte der ökologischen Bewegungen in Deutschland* [*Ecopax and Anarchy: A History of Ecological Movements in Germany*] (Munich, 1986). Linse's fascinating study concentrates particularly on the socialist and anarchist precursors of modern green movements. Jost

Hermand's *Grüne Utopien* is an impressive exercise in scholarship to which I am much indebted, and which traces the development in ecological consciousness in Germany from the eighteenth century onwards. I am likewise indebted to Raymond H. Dominick III's detailed and essential history of conservation movements in Germany: *The Environmental Movement in Germany: Prophets and Pioneers, 1871–1971* (Bloomington and Indianapolis, 1992). In addition to Brüggemeier and Rommelspacher (eds.), *Besiegte Natur*, which *inter alia* contains valuable contributions on the history of energy use as well as of air and water pollution, some other volumes relevant to the German dimension are: Klaus Bergmann, *Agrarromantik und Großstadtfeindschaft* [*Rural Romanticism and Hostility to Cities*] (Meisenheim am Glan, 1970); Rolf Peter Sieferle, *Fortschrittsfeinde? Opposition gegen Technik und Industrie von der Romantik bis zur Gegenwart* [*Enemies of Progress? Opposition to Technology and Industry from the Romantics to the Present*] (Munich, 1984); Ludwig Trepl, *Geschichte der Ökologie vom 17. Jahrhundert bis zur Gegenwart* [*History of Ecology from the Seventeenth Century to the Present*] (Frankfurt am Main, 1987); Anna Bramwell, *Ecology in the Twentieth Century: A History* (New Haven and London, 1989); Peter C. Mayer-Tasch, *Natur denken. Eine Genealogie der ökologischen Idee* [*Thinking Nature: A Genealogy of the Ecological Idea*] (Frankfurt am Main, 1991); Jost Hermand (ed.), *Mit den Bäumen sterben die Menschen. Zur Kulturgeschichte der Ökologie* [*When the Trees Die, So Do the People: On the Cultural History of Ecology*] (Cologne, Weimar and Vienna, 1993); Werner Abelshauser (ed.), *Umweltgeschichte. Umweltverträgliches Wirtschaften in historischer Perspektive* [*Environmental History: Environmentally Sustainable Economic Action in Historical Perspective*] (Göttingen, 1994).

[3.] See, for example, Anna Bramwell, *The Fading of the Greens: The Decline of Environmental Politics in the West* (New Haven and London, 1994). Ingolfur Blühdorn describes the contemporary German environmental movement as 'an irritating mixture of success and crisis' in 'Campaigning for nature: environmental pressure groups in Germany and generational change in the ecology movement', in Ingolfur Blühdorn, Frank Krause and Thomas Scharf (eds.), *The Green Agenda: Environmental Politics and Policy in Germany* (Keele, 1995), 167–220, here 169.

[4.] Dominick, for instance, classifies his analysis of the origins of environmental movements under those two broad headings. See *Environmental Movement*, 8–41.

[5.] See Joachim Radkau, 'Was ist Umweltgeschichte?' ['What is environmental history?'], in Abelshauser, *Umweltgeschichte*, 11–28, here 15. This is essentially a restatement of anthropocentrism in its 'weak sense', as defined by Andrew Dobson, *Green Political Thought*, 2nd edn. (London, 1996), 61.

6. Dobson, *Green Political Thought*, 14–38 and *passim*. The distinction Dobson draws has been critically evaluated since the first edition of his book (1990); see especially Tim Hayward, *Ecological Thought: An Introduction* (Cambridge, 1994), 187–99. Ingolfur Blühdorn draws on Goodin to stress the continuity between the categories of conservationism, environmentalism and ecologism, as part of an argument to explain a paradigm shift in green politics. See Blühdorn, 'Campaigning for nature', 203, and Robert E. Goodin, *Green Political Theory* (Cambridge, 1992). A threefold typology of conservationism, environmentalism and ecologism was suggested by Dieter Rucht in 1989, though in much less detail than Dobson. See Rucht, 'Environmental movement organizations in West Germany and France: structure and interorganizational relations', *International Social Movement Research*, 2 (1989), 61–94 (see p. 64).

7. The form of government necessary to an ecologically sustainable society is the subject of much debate; there is something to be said for the view that authoritarian models are not truly ecological. See, for example, the section on 'Green politics: the state and democracy', in Andrew Dobson and Paul Lucardie (eds.), *The Politics of Nature. Explorations in Green Political Theory* (London and New York, 1993).

8. Anna Bramwell traces the origin of the 'ecological box' as she puts it, to about 1880. It may be true that environmentalism existed at that date in rudimentary form, but it would be very difficult to prove that ecologism in the modern sense, with all that it implies, existed any earlier than the later 1950s. See Bramwell, *Ecology in the Twentieth Century*, 13–21.

9. See especially Dominick, *Environmental Movement*.

10. Dobson, *Green Political Thought*, 35.

11. Mayer-Tasch, *Natur denken*, 32–3.

12. Ibid., 65–6. Plato was also a remarkably prescient observer of ecological disruption: Wall quotes an excerpt showing his understanding of soil erosion through overgrazing (*Green History*, 36–7). Jost Hermand is thus wrong in stating that 'a concern for the environment threatened by man' is missing in Plato's writings (Hermand, *Grüne Utopien*, 21).

13. Mayer-Tasch, *Natur denken*, 64 and 72.

14. Ibid., 91.

15. Hermand, *Grüne Utopien*, 26 and 36–8.

16. Compare Wall, *Green History*, 3.

17. Compare ibid., 78, on Gaia. J. E. Lovelock's Gaia is not a myth, but a scientific hypothesis, however controversial.

18. See *Goethes Werke*, ed. Benno von Wiese and Erich Trunz (Hamburg, 1958), XIV, 145–53, and cf. Hermand, *Grüne Utopien*, 50–9

19. Eric A. Blackall, *The Novels of the German Romantics* (Ithaca and London, 1983), 145.

20. For considerations of mythical and ecological apocalypse, see Axel

Goodbody in ch. 9 and Tom Cheesman in ch.17 of this volume.

[21.] An excellent history of the science of ecology, as distinct from ecological politics, can be found in Donald Worster, *Nature's Economy: A History of Ecological Ideas*, 2nd edn. (Cambridge, 1994).

[22.] Ibid., 135.

[23.] Alexander von Humboldt, *Ansichten der Natur*, ed. Adolf Meyer-Abich (Stuttgart, 1969), 56.

[24] See Hermand, *Grüne Utopien*, 45–6.

[25.] Percy Bysshe Shelley, *A Vindication of Natural Diet* (London, 1813), 20–3, quoted in Wall, *Green History*, 130–1.

[26.] See Worster, *Nature's Economy*, 133.

[27.] See Dominick, *Environmental Movement*, 18–19 and 36.

[28.] See Radkau, 'Was ist Umweltgeschichte?', 26. Radkau cites Andrée Corvol, *L'Homme aux bois. Histoire des relations de l'homme et de la forêt XVIIe–XXe siècle* (Paris, 1987).

[29.] My translation of Ernst Moritz Arndt, 'Ein Wort über die Pflegung und Erhaltung der Forsten und Bauern im Sinne einer höheren, d. h. menschlichen Gesetzgebung' ['A word on the care and preservation of the forests and the peasants in the sense of a higher, i.e. humane law'], *Der Wächter* (1815), 384ff., quoted in Hermand, *Grüne Utopien*, 44–5.

[30.] See Wall, *Green History*, 126–9 and *passim*.

[31.] The history of this affinity is traced in Simon Schama, *Landscape and Memory* (London, 1995).

[32.] Wilhelm Heinrich Riehl, *Die Naturgeschichte des Volkes als Grundlage einer deutschen Sozial-Politik*, I, *Land und Leute*, 2nd edn. (Stuttgart and Augsburg, 1855), 44.

[33.] The notion that landscape can explain national character was a long-standing one, shared, among others by Herder, Moser, and, indeed, Alexander von Humboldt.

[34.] Riehl, *Naturgeschichte*, 44.

[35.] Ibid., 45.

[36.] Ibid., 46.

[37.] Ibid., 47.

[38.] Bramwell, in *Ecology in the Twentieth Century*, gives a detailed account of the supposed provenance of the term (253, note 2).

[39.] Daniel Gasman, *The Scientific Origins of National Socialism: Social Darwinism in Ernst Haeckel and the German Monist League* (London, New York, 1971), 6.

[40.] Ibid., 34–5. Gasman's main concern is to argue that Haeckel's social Darwinism made him into a precursor of Nazism. Kelly counters that this 'ignores the rational, liberal, and humanitarian side of monism': Alfred Kelly, *The Descent of Darwin: The Popularization of Darwinism in Germany, 1860–1914* (Chapel Hill, 1981), 120. Bramwell concurs (see

Ecology in the Twentieth Century, 50–1).

41. Ernst Haeckel, *Eternity: World War Thoughts on Life and Death, Religion, and the Theory of Evolution* (New York, 1916), 14, quoted in Gasman, *Scientific Origins of National Socialism*, 33.

42. In Germany, Bölsche became 'probably the greatest science popularizer of all time; and as the author of dozens of best-selling books and hundreds of articles, his name was a household word to millions'. Kelly, *The Descent of Darwin*, 36.

43. See, for example, Dominick, *Environmental Movement*, 1, or Jefferies, in this volume, 42.

44. For example: 'On at least three occasions, in 1715, 1721, and again in 1778, Stuttgarters who lived alongside the unhappy Nesenbach petitioned their government to correct the silting and pollution of that stream.' Dominick, *Environmental Movement*, 42. Incidentally, the topic under discussion here is reaction to environmental depredations, not the depredations themselves. A number of studies show how human beings have affected the environment since neolithic times: see, for example, excerpts quoted in Wall, *Green History*, 33–5, or Clive Ponting, *A Green History of the World* (London, 1991).

45. Sieferle, *Fortschrittsfeinde*, 58–60.

46. Arne Andersen and Franz-Josef Brüggemeier, 'Gase, Rauch und Saurer Regen' ['Gases, smoke and acid rain'], in Brüggemeier and Rommelspacher (eds.), *Besiegte Natur*, 64–85.

47. Thomas Rommelspacher, 'Das natürliche Recht auf Wasserverschmutzung' ['The natural right to pollute water'], in Brüggemeier and Rommelspacher (eds.), *Besiegte Natur*, 42–63, here 44.

48. See ibid.

49. Matters were complicated by the fact that scientists could not agree on the cause of disease in towns, especially cholera. For a detailed account, see Richard Evans, *Death in Hamburg: Society and Politics in the Cholera Years, 1830–1910* (Oxford, 1987).

50. See Andreas Knaut, 'Die Anfänge des staatlichen Naturschutzes. Die frühe regierungsamtliche Organisation des Natur- und Landschaftsschutzes in Preußen, Bayern und Württemberg' ['The beginnings of state-sponsored nature conservation'], in Abelshauser (ed.), *Umweltgeschichte*, 143–62, or Dominick, *Environmental Movement*, for details.

51. See Hermand, *Grüne Utopien*, 46ff., and Dominick, *Environmental Movement*, 57.

52. See Dominick, *Environmental Movement*, 47–8, 53–4 and 179–81.

53. It is worth noting that this very broad heading does not necessarily imply a pro-capitalist stance, and that not all members of mainly nationalist movements can easily be categorized on the political right.

54. Ernst Rudorff, 'Über das Verhältnis des modernen Lebens zur Natur',

quoted in Arne Andersen, 'Heimatschutz: Die bürgerliche Natur-schutzbewegung' ['Protection of the homeland: the bourgeois nature-protection movement'], in Brüggemeier and Rommelspacher (eds.), *Besiegte Natur*, 143–57, here 143.

55. See Andersen, 'Heimatschutz', 144–5, and Hermand, *Grüne Utopien*, 85–7.

56. Dominick, *Environmental Movement*, 51.

57. Ibid., 53.

58 Conwentz, quoted in Andersen, 'Heimatschutz', 147.

59. On Schultze-Naumburg, see Jefferies in ch. 2 of this volume.

60. See Jefferies in this volume, 48, Dominick, *Environmental Movement*, 57, and Andersen, 'Heimatschutz', 149.

61. Andersen, 'Heimatschutz', 150.

62. William Rollins, '"Bund Heimatschutz": Zur Integration von Ästhetik und Ökologie' ['"Bund Heimatschutz". On the integration of aesthetics and ecology'], in Hermand (ed.), *Mit den Bäumen sterben die Menschen*, 149–82.

63. Ibid., 154.

64. Blühdorn, 'Campaigning for nature', 205.

65. See Pepper, *Roots of Modern Environmentalism*, 151.

66. Bramwell, *Ecology in the Twentieth Century*, 31–6.

67. Hermand, *Grüne Utopien*, 76–7. Of course, Marx and Engels are important sources for *modern* green thought. See, for example, Pepper, *Roots of Modern Environmentalism*.

68. Bebel asserts gender equality in all ways throughout: 'There can be no liberty for humankind without the social independence and equality of the sexes.' August Bebel, *Die Frau und der Sozialismus*, 31st edn. (Stuttgart, 1900), 7.

69. Ibid., 352.

70. Ibid., 354.

71. Compare Hermand, *Grüne Utopien*, 79–81. Dominick describes Bebel's views on the environment as 'managerial rather than preservation-ist', *Environmental Movement*, 63.

72. 'In the socialist society increased production will be to the advantage of all'. Bebel, *Die Frau und der Sozialismus*, 353.

73. See ibid., 355.

74. See Dominick, *Environmental Movement*, 125–30 and *passim*.

75. See Rolf Peter Sieferle, 'Energie', in Brüggemeier and Rommelspacher (eds.), *Besiegte Natur*, 20–41, here 35. In Britain, this question had been investigated by parliamentary commissions in the first half of the nineteenth century.

76. See Bebel, *Die Frau und der Sozialismus*, 353.

77. Ibid., 354–6.

78. Ibid., 356.

79. Ibid.

80. This account is drawn largely from Jochen Zimmer, 'Soziales Wandern. Zur proletarischen Naturaneignung' ['Social hiking: on the proletarian appropriation of nature'], in Brüggemeier and Rommelspacher (eds.), *Besiegte Natur*, 158–67.

81. Ibid., 163.

82. For an account of back-to-nature movements in Britain, see Peter C. Gould, *Early Green Politics: Back to Nature, Back to the Land, and Socialism in Britain 1880–1900* (Brighton and New York, 1988).

83. See Hermand, *Grüne Utopien*, 93.

84. For more on Steiner, see Frank Finlay's contribution in ch.14 of this volume, especially 254–5.

85. See Ulrich Linse (ed.), *Zurück, o Mensch, zur Mutter Erde. Landkommunen in Deutschland, 1890–1933* (Munich, 1983) for more details.

86. See Ben Macintyre, *Forgotten Fatherland: The Search for Elisabeth Nietzsche* (London, 1992).

87. They include Hesse, D. H. Lawrence, Kafka, Jung, for example. See Michael Green, *Mountain of Truth: The Counterculture Begins. Ascona, 1900–1920* (Hanover, NH, and London, 1986).

88. Ibid., 120–2.

89. Ibid., 123.

90. Dominick excludes such groups entirely from *The Environmental Movement in Germany*.

91. Ludwig Klages, *Mensch und Erde: zehn Abhandlungen* (Stuttgart, 1956), 1–25, here 21.

92. See Linse, *Ökopax*, 81.

93 Linse, *Ökopax*, 95.

94. Ibid., 95.

95. Ibid., 83.

96. Ibid., 86.

97. Ibid., 116.

98. Ibid., 119.

99. Dominick, *Environmental Movement*, 88.

100. Dominick cites Eugen Diesel as an example of a Heimatschützer who rejected anti-Semitism and xenophobic nationalism. See *Environmental Movement*, 97.

101. Paul Schultze-Naumburg, 'Die Gestaltung der Landschaft', in *Der deutsche Heimatschutz. Ein Rückblick und Ausblick* (Munich, 1930), 11–17 (here, 14), quoted in Sieferle, *Fortschrittsfeinde*, 201.

102. Sieferle, *Fortschrittsfeinde*, 201.

103. Dominick, *Environmental Movement*, 96.

[104.] The legislation was printed as an appendix to post-1933 editions of R. Walther Darré, *Das Bauerntum als Lebensquell der nordischen Rasse* (Munich, 1935), 469.

[105.] The *Lebensraum* policy contradicted 'blood and soil', since it would by definition break the link with German soil.

[106.] See Sieferle, *Fortschrittsfeinde*, 217-8.

[107.] See Anna Bramwell, *Blood and Soil: Walther Darré and Hitler's 'Green Party'*, (Abbotsbrook, 1985).

[108.] Much of this part of the account is drawn from Dominick, *Environmental Movement*, 99-108.

[109.] Sieferle, *Fortschrittsfeinde*, 213-15.

[110.] See Dominick, *Environmental Movement*, 119-21.

[111.] Dominick charts the composite membership of four leading groups. *Environmental Movement*, 122.

[112.] Ibid., 161-2.

[113.] Ibid., 157. See Bodo Manstein, *Im Würgegriff des Fortschritts* [*In the Stranglehold of Progress*] (Frankfurt am Main, 1961).

[114.] For details, see Dominick, *Environmental Movement*, 160-8. There was also a spate of apocalyptic literature during this period; see Axel Goodbody in ch. 9 of this volume.

[115.] See Hermand, *Grüne Utopien*, 122-23, and Dominick, *Environmental Movement*, 148-58.

[116.] Günter Schwab, *Der Tanz mit dem Teufel. Ein abenteuerliches Interview*, 15th edn. (Hamelin, 1991) (first published 1958).

[117.] Ibid., 419.

[118.] Ibid., 452.

[119.] Ibid., 448.

[120.] It is worth noting, however, that Schwab founded and led probably the biggest environmentalist organization in the 1960s. His *Weltbund zum Schutz des Lebens* [World League for the Protection of Life] eventually numbered over one million members in West Germany (see Dominick, *Environmental Movement*, 156.)

[121.] See Hermand, *Grüne Utopien*, 129.

[122.] See Dominick, *Environmental Movement*, 169-79.

[123.] Ibid., 184.

[124.] In addition to Blühdorn, Krause and Scharf (eds.), *The Green Agenda*, which is an excellent, wide-ranging retrospective and snapshot of German green policy and politics in the mid-1990s, see also: Charlene Spretnak and Fritjof Capra, *Green Politics: The Global Promise* (London, 1984); Elim Papadakis, *The Green Movement in West Germany* (London, 1984); Manuel Dittmers, *The Green Party in West Germany: Who are they? And What Do they Really Want?* (Buckingham, 1988); Werner Hülsberg, *The German Greens: A Social and Political Profile* (London, 1988); Eva Kolinsky (ed.),

The Greens in West Germany: Organisation and Policy Making (Oxford, 1989); Thomas Scharf, *The German Greens: Challenging the Consensus* (Oxford and Providence, 1994). See also part II, 'Contemporary Trends', in this volume.

125. The text was reissued under the title *Mensch und Erde. Ein Denkanstoß* (Bonn, 1980); see Linse, *Ökopax*, 60.

126. Ernst Bloch's *Das Prinzip Hoffnung* or the Frankfurt School are more common sources (see, for example, Ingolfur Blühdorn in ch. 5 of this volume).

127. Anna Bramwell, 'Ricardo Walther Darré – was this man "Father of the Greens"?', *History Today*, 34 (1984), 7–13.

128. For more detail on this, see Alison Statham in ch. 7 of this volume.

129. See Blühdorn, 'Campaigning for nature'.

2 • Heimatschutz : Environmental Activism in Wilhelmine Germany

Matthew Jefferies

The historic roots of German ecologism are tangled and muddy. Those who seek to trace a linear evolution of green thought through the work of German poets and philosophers face many frustrations, just as those who posit a special relationship between the German people and their forests, stretching from Arminius to Beuys, struggle to substantiate what is undoubtedly a stimulating proposition.[1] By comparison, the search for the origins of organized environmental activism in Germany may appear relatively straightforward. It is invariably suggested that they lie in the turbulent decades of rapid industrial and urban growth between unification in 1871 and the outbreak of war in 1914.[2]

Certainly these were years in which many key milestones in the history of German environmentalism were passed. It was in 1904, for instance, that the Bund Heimatschutz [League for the Protection of the Homeland], the first mass movement in the history of German environmentalism, was established at a packed public meeting in Dresden. The 1900s also witnessed: the first mass protest against a large engineering project in an environmentally sensitive location (namely, the unsuccessful campaign in 1903–4 to stop the Laufenburg rapids in Baden from being lost as part of a hydroelectric power scheme); the first organized attempts to create national parks in Germany (the Verein Naturschutzpark [Society for Nature Reserves], founded in 1909, purchased some 4,000 hectares of land on Lüneburg Heath between 1910 and 1920); the first campaigns against the overdevelopment of areas of outstanding natural beauty, such as the Alps, for tourism; and finally, the first town and country planning legislation to offer real protection for the German landscape, the Prussian 'Law against the Disfigurement of Places and Outstanding Landscapes' of 1907, and similar legislation in many other states.

However, the character of – and motivation for – these green landmarks are a good deal less straightforward than one might imagine. As will become apparent during the course of this chapter, the prevailing attitude of Wilhelmine environmentalists to the natural world was aesthetic and sentimental rather than ecological. This had less to do with the extent of scientific knowledge in turn-of-the-century Germany than with the nature of the Wilhelmine environmental movement itself. For although these years produced a well-informed scientific discourse on the threat posed by industrial pollution, alongside pioneering works of ecological theory from the likes of Haeckel and Ostwald, neither scientists nor monists featured prominently in popular environmental campaigns at the turn of the century. This chapter seeks an explanation for this paradox in the particular origins of the Wilhelmine environmental movement, which emerged as part of a wider cultural reform tendency, whose concerns were primarily aesthetic in nature. Nowhere is this better illustrated than in the character of the first chairman of the Bund Heimatschutz, Paul Schultze-Naumburg (1869–1948), whose writings on the environment provide the main focus of the following pages.

Schultze-Naumburg was an artist by training and temperament, who had settled in Munich, the artistic capital of Imperial Germany, in the early 1890s.[3] Here he had established a private art school with his first wife Ernestine. The 1890s were a turbulent time in both the applied and fine arts in Germany, as a new generation of artists, designers and architects sought to throw off the shackles of academic historicism, in the search for an art that would be closer to the people, and more in touch with the spirit of the age. Later, the contradictions and contrasts between those who looked for radically new solutions, and those who sought to re-establish continuity with the values of the past, would become all too apparent, but for the time being, such distinctions were blurred by a shared contempt for the excesses of the *Gründerzeit* [founding years of the empire] and a common vocabulary which placed emphasis on honesty and simplicity. Schultze-Naumburg duly became involved in every significant reform movement of the day, from the Munich and Berlin Secessions to the German Werkbund, as well as the Heimatschutz activities under consideration here.

From an early age, Schultze-Naumburg sought opportunities to express his views in print. The didactic tone that became such a feature of his best-known writings was apparent from the very beginning.[4] In 1901 he published one of his most widely-read pamphlets, on *The Culture of the Female Body as the Basis of Women's Clothing*, which offered 'a scientific and aesthetic assessment of the female body'.[5] Essentially a fierce attack on contemporary fashion, and particularly the corset which Schultze-Naumburg believed was crippling women's bodies, the essay also had a political subtext, because fashion in Germany was still largely dictated by Paris.

Since the mid-1890s Schultze-Naumburg had also been contributing reviews and essays to Ferdinand Avenarius's influential journal *Der Kunstwart*, which at this time was emerging as the principal mouthpiece of the reform movement.[6] Avenarius, who had founded the journal in 1887, soon recognized a kindred spirit in Schultze-Naumburg and appointed him chief art correspondent. It was Schultze-Naumburg's *Kunstwart* essays which formed the basis of his most substantial literary undertaking, the so-called *Kulturarbeiten* [*Cultural Works*], of which no fewer than nine volumes appeared between 1901 and 1917. The starting-point for the series was the conviction that every member of the public had an influence on the environment and that all individuals could and should undertake their own *Kulturarbeit*. Schultze-Naumburg was convinced that by educating the layman in areas such as architecture, town planning and conservation, it would be possible to reverse the damaging trends of the previous decades. As he put it, 'Its purpose is to work against the horrific devastation of our country in every field of visual culture.'[7]

Thus each reasonably priced volume was written in an uncomplicated, accessible and occasionally even humorous style, in large print and with numerous illustrations. The pages of photographs, which outnumbered the pages of text, were laid out as 'example' and 'counter-example', with one photograph per page. This simple juxtaposition of 'good' and 'bad' proved an enduring and highly effective device, which was much imitated over the following years. Most of the 2,500 photographs used in the *Kulturarbeiten* were taken by Schultze-Naumburg himself, and were works of art in their own right. The visual emphasis in general eschewed the great, the important and the dramatic, in favour of the humble and the anonymous. The significance of the *Kulturarbeiten* was recognized by the respected urban historian Lewis Mumford, who described the series as 'a work of fundamental importance upon the artful and orderly transformation of the environment by man. One of the original monuments of its generation'.[8]

Schultze-Naumburg's views on architecture, as revealed in volume 1 of the *Kulturarbeiten*, were to remain remarkably consistent throughout his life. His contempt for the practice of designing contemporary buildings in inappropriate historical styles – in particular the willingness of provincial master-builders to pluck styles at random from the pattern-books – put him at the forefront of the architectural-reform movement. Ironically, his ideas also had a formative influence on a number of the modern architects who were to become his sworn enemies in the 1920s. In place of textbook historicism, or indeed the rootless experimentation with 'new' styles, which he found equally repellent, Schultze-Naumburg consistently advocated an architecture which employed local materials and continued in the spirit of local vernacular traditions, a line of continuity which he believed had been severed earlier in the nineteenth century, with disastrous consequences for the German landscape. His particular enthusiasm was for an architectural tradition embodied by Goethe's 'garden house' in Weimar: timeless, simple and in harmony with its surroundings.

The popular appeal of the *Kulturarbeiten* and Schultze-Naumburg's other writings opened up new career opportunities; most notably the chance to design buildings of his own. Despite the absence of any formal architectural training, Schultze-Naumburg began to receive numerous commissions, mostly from wealthy patrons for substantial country and suburban homes. He completed over one hundred major projects between 1904 and 1930, including additions to his own home and craft workshops at Saaleck in the picturesque Saale valley. As one would expect, his architecture was solid and worthy, rather than innovative or inspired; it never received the same acclaim as his writing, even from his admirers. However, one thing that even his harshest critics frequently praised was his ability to make the most of a building's location. His biographer Norbert Borrmann explained this in the following terms: 'The architect Schultze-Naumburg designed buildings which form an integrated whole with the landscape, thereby recreating the subjects which had been lost to the painter Schultze-Naumburg.'[9]

It was this concern with the *context* of architecture, the relationship of the built environment to the natural world, which became the dominant theme in Schultze-Naumburg's writings, and which first brought him into contact with green issues. The volumes of the *Kulturarbeiten* which deal with environmental questions in most depth (volumes 7–9) had as their unifying theme man's impact on the landscape, which Schultze-Naumburg examined under six main headings: tracks and roads; forestry and agriculture; quarrying and mining; waterways; industry and railways; and settlements. It was only in the course of the nineteenth century, with the onset of industrialization, he argued, that such activities had developed a momentum of their own, and that the ability to maintain harmony with nature had been lost. The search for short-term profit had become paramount, and the long-term costs invariably forgotten.

To illustrate this, he cited the example of farmers who had been encouraged to remove hedgerows to boost their productive area, but in so doing had destroyed a valuable wildlife habitat. The insect population, previously kept in check by the birds which had nested in the bushes, promptly multiplied, and the farmers were forced to use artificial methods to combat the resulting plague. Any profit made from the removal of the hedges was thus spent on insecticides, which made little sense financially, and could also lead to unforeseen ecological problems. As Schultze-Naumburg put it: 'This is what usually happens, when the oh-so-clever human removes a little cog from the great mechanism, because it irritates him and because he doesn't realize that the whole mechanism will then not work'.[10]

This did not mean, however, that he believed the advance of technology could or should be halted. He did not advocate a return to a Ruskinesque pre-industrial idyll, nor a fundamental rethink of the capitalist system. In fact, many of the attitudes expressed by Schultze-Naumburg in his pre-war writing would raise few eyebrows today. He was an advocate of more manageable cities, in which the citizen should always be able to reach the country without difficulty; he attacked the planners' reverence for the needs of traffic;[11] and he attacked industrial firms for turning Germany's rivers into 'stinking sewers'.

In contrast to those late nineteenth-century cultural pessimists, who could see only decadence and decay on the horizon, his writing was characterized by a touching, if rather naive, optimism that his educational efforts could really make a difference. What is more, he was quite prepared to back up his words with actions, becoming an active campaigner and lobbyist in his role as chairman of the Bund

Heimatschutz, founded in March 1904.[12]

Although the origins of the Heimatschutz idea can be traced back to the Romantic era and beyond, the more immediate roots of the Bund Heimatschutz lay in the late nineteenth century, when a number of writers had begun to express concern for the passing of diversity in the German landscape. The combined effects of industrialization, rapid urbanization and political unification were, it was suggested, wiping out centuries of culture and tradition. For Ernst Rudorff, the Berlin professor of music who wrote the influential essay 'On the relationship of modern life to nature' in 1880 and first coined the term *Heimatschutz* in 1897, the natural world had a particular significance for the Germans:

> The very roots of the German character lie in a profound and intimate feeling for nature. It is that which our ancestors found entrancing in Wodin's holy oak grove; that lives in the sagas of the Middle Ages in the figures of Melusina or the Sleeping Beauty; that crops up in the songs of Walther von der Vogelweide and then, to a new and undreamt of extent, in Goethe or Eichendorff's poetry; and, finally, that bursts from the most particular revelation of German genius, our marvellous music. It is the same keynote, the same deep pull of the soul towards the wonderful and unfathomable secrets of nature, speaking through all these expressions of the national disposition.[13]

For the likes of Rudorff, the principal motivation for Heimatschutz came from a fear and loathing of the modern metropolis, its mass culture and, above all, its organized working class. His reactionary tone was unmistakable. As I have stressed elsewhere, however, there was another, more pragmatic strand within the Heimatschutz movement, more concerned with practical conservation than political posturing, and sharing much in common with the members of the British National Trust, which had been founded less than a decade earlier.[14] The choice of Schultze-Naumburg as the League's first chairman was a logical one: he was close friends with Rudorff and in general shared his conservative outlook, but he was less élitist, and his efforts to popularize the Heimatschutz message had already succeeded in gaining a large following of predominantly middle-class Germans, anxious to make their own practical contribution to environmental improvement.

According to its founding statutes, the League's activities were to be concentrated in six areas:

- the preservation of monuments
- the promotion of vernacular building styles
- the conservation of the landscape
- the protection of indigenous species, flora and fauna, and geology
- the promotion of traditional arts and crafts
- the preservation of country customs, festivals and costumes

It soon became evident, however, that the League's principal concern was for the landscape. The aforementioned struggle to save the rapids on the Rhine at Laufenburg, between Konstanz and Basle, which were threatened by a hydroelectric power project, was the new organization's first major challenge and set the tone for subsequent campaigns. It also prompted a long series of articles by Schultze-Naumburg on the theme of Heimatschutz in *Der Kunstwart*, and provoked one of his most forthright ecological statements:

> A time will come when it will be recognized that man does not live on horsepower and machines alone; that there are other possessions which he cannot and would not want to do without . . . For, if man were to extract everything which can be extracted by his technology, he would come to realize that the resulting easy life on a disfigured Earth is actually no longer worth living; that we would have grabbed everything that our planet has to offer, but in so doing we would have destroyed it and therefore ourselves as well. It is up to each and every one of us to do our bit to bring about a change before it is too late, everywhere and for ever.[15]

The campaign to save the rapids failed, despite the support of some of the most prominent figures in Wilhelmine intellectual life, including Friedrich Naumann, Werner Sombart and Max Weber. Nevertheless, the high-profile campaign helped to spread the Heimatschutz message, and Schultze-Naumburg's prophetic words were to have a lasting impact. Amongst the many people to be impressed was Walther Darré, later Hitler's agriculture minister, who used the very same quotation to introduce his most famous work, *Neuadel aus Blut und Boden* [*A New Aristocracy from Blood and Soil*], which was not only dedicated to Schultze-Naumburg, but written in his house during a prolonged stay at Saaleck in the late 1920s.

Schultze-Naumburg returned to the subject of hydroelectric power stations and dams in a 1906 article for *Der Kunstwart*. He made it clear that whilst he was not against either in principle, each project should be carefully evaluated for its costs and benefits, since each time a valley was deliberately flooded, a great deal was at stake. As he pointed out:

> If Cologne Cathedral stood in that valley and one wanted to drown it in water in return for so or so many dollars, no doubt a storm of protest would be raised. I simply cannot see why the masterpieces of God's natural world should have less right to existence than man-made works of art.[16]

Schultze-Naumburg's last word on the Laufenburg issue was to express hope that the power station would be erected 'in the most attractive form possible'. This was a theme he would address on numerous occasions. Where the construction of industrial or other technical installations was unavoidable, he argued, they should always be designed to blend in with their surroundings. This did not mean, however, that their true function should be disguised. Indeed, as he put it, 'In most cases it is not even the technical part of our modern industrial installations which is such a slap in the face, but rather it is the wretchedly stupid "finery" that is put on the thing'.[17] Attempting to influence the quality of industrial architecture subsequently became one of the League's main areas of activity, with the organization sponsoring attempts to find simple, functional and aesthetically pleasing designs for factories, railway stations and electricity substations in environmentally sensitive locations.

The League also attempted to secure official backing for its efforts, and to influence legislation at both a local and a national level. In this it could point to some notable successes: for instance, Schultze-Naumburg's 1905 pamphlet for the League, 'The disfigurement of our country', was purchased from public funds in a number of districts, and recommended by ministries in Prussia, Bavaria and Württemberg amongst others. Many of its 50,000 copies were distributed in the aftermath of the 1907 Prussian 'Law against the Disfigurement of Places and Outstanding Landscapes', the so-called *Verunstaltungsgesetz,* which was emulated in most other German states before the First World War. Under the Prussian law, which was in advance of anything seen in Britain until the 1932 Town and Country Planning Act, the authorities gained the power to refuse planning permission for any building which 'grossly disfigured' the landscape. At the same time, top civil servants in each of Prussia's provinces were asked to draw up lists of areas of outstanding natural beauty, to be safeguarded against development.[18]

Even so, one should not exaggerate the effectiveness of the Heimatschützer in combating the many threats faced by the German environment. In 1906 Schultze-Naumburg had written:

> A tree, from which only a couple of branches jut up into the air like virile shoots whilst all the others are wasting away and dying, is not a healthy organism, for which a long life can be predicted. Our current culture resembles just such a tree. If we want to do our bit to bring this danger under control, we must not be like the bad doctor and only treat the external symptoms of the disease. Rather we must get to the bottom of things and seek to strengthen those cultural ideas which are suited to giving humanity a higher and more harmonious view of life.[19]

Yet one could argue that a 'bad doctor' is precisely what Schultze-Naumburg himself resembled, for his attempts to reveal the underlying causes of environmental damage were not on the whole characterized by the same dogged determination which underpinned his admirable efforts to educate the public on aesthetic issues. In his pre-war writings Schultze-Naumburg sought an explanation for environmental decay in the ethical and moral shortcomings of modern man; later, he would come to view the pseudo-science of racial hygiene and the 'degeneration' of the German *Volk,* as the prime cause of environmental damage. Neither approach was likely to deliver sensible solutions to the genuine problems he had recognized.

Just as both Schultze-Naumburg in person, and the Heimatschützer in general, were more successful at recognizing the superficial symptoms of environmental destruction than its causes, so the insidious but often invisible impact of pollution on the German landscape also – for the most part – evaded their gaze. The predominance of aesthetes in the movement ensured that their approach to environmental issues was based on contemplation and empathy rather than reflection or scientific analysis. Organizations like the Verein gegen die Verunreinigung der Flüsse, des Bodens und der Luft [Union against the Pollution of the Rivers, Soil and Air], which had been active since 1877, and journals such as *Rauch und Staub [Smoke and Dust]*, which frequently discussed the damaging effects of industrial pollution, were ignored by the Heimatschützer.

In fact, as a number of recent studies have pointed out, much more was known about the ecological problems in turn-of-the-century Germany than the Heimatschutz discourse would suggest.[20] Moreover, their preoccupation with the aesthetics of the environment could sometimes blind them to reality. The following quotation comes from Eugen Gradmann, a state conservation officer in Stuttgart, writing in a Heimatschutz pamphlet:

> It is rather comforting that even when the air is clouded by smoke, smog and fog, the results can be most picturesque . . . The light is broken up, giving the colourful diversity of things a more unified tone; a vivid atmosphere, which can also arouse in us a spiritual mood . . . [21]

There was also, of course, a particular irony in the fact that it was a hydroelectric power scheme which had so galvanized the Wilhelmine Heimatschützer. This comparatively 'green' method of energy production had come into vogue at the turn of the century because of fears surrounding the dwindling reserves of fossil energy – an aspect of ecology which Schultze-Naumburg never addressed.

It would be unfair, however, to attack Schultze-Naumburg and the Wilhelmine Heimatschützer on the grounds of what they were not. From the beginning, environmentalism and ecologism took different paths. The pragmatic, problem-solving approach pursued by the Bund Heimatschutz clearly followed the former rather than the latter route.

Schultze-Naumburg's tireless efforts to communicate the importance of the landscape continued after the war, with books like *Vom Verstehen und Genießen der Landschaft* [*Understanding and Enjoying the Landscape*], a handbook for young hikers published in 1924, but by then his personal fortunes had taken a turn for the worse. Like most German architects, he could build little between 1915 and 1923, and was reduced to writing 'begging letters' to friends and former patrons. In November 1920, for instance, he wrote to the industrialist Gustav Krupp von Bohlen und Halbach asking: 'May I put the old apprentice's question to you: don't you have any work for me? My once overcrowded offices are now empty and desolate.'[22] Like most of his requests for work, this met with a negative response, and business remained slow even after the recovery of building activity in the mid-1920s. To his credit, Schultze-Naumburg did not compromise his aesthetic principles in an attempt to compete with the radical visions of his youthful competitors, but the stagnation in his career no doubt made him more receptive to the easy answers and ready scapegoats offered by racial theory, the dominant theme of all his later works.

Thus, when Wilhelm Frick became the first member of the National Socialist German Workers' Party to attain ministerial office, as Minister of the Interior and Education in the Thuringian coalition government of 1930, he moved quickly to appoint Schultze-Naumburg as head of a reformed school of art and architecture in Weimar, formerly home to the controversial Bauhaus.[23] Within months, Schultze-Naumburg became the principal spokesman on artistic matters for the Nazi Kampfbund für deutsche Kultur [Militant League for German Culture], which republished many of his writings and sponsored him on lecture tours of the Reich, accompanied by a full contingent of SA storm-troopers and frequent public disorder. At a time of life when he might reasonably have been contemplating a quiet retirement, Schultze-Naumburg was thrust before the flashbulbs and flying fists, undergoing a personal radicalization every bit as dramatic as that experienced by the German body politic as a whole.

It is as a Nazi sympathizer and racist ideologue that Schultze-Naumburg is now best remembered. His own architectural output and, more surprisingly, his substantial body of pre-war writings have in contrast been largely forgotten. Nevertheless, the work of Schultze-Naumburg deserves a place in any study of green thought in German culture, for its central role in galvanizing the Heimatschutz movement, its impact on planning legislation, and its influence on a generation of young Germans. Of course, the last is notoriously difficult to measure, but Julius Posener, the doyen of German architectural historians, has said that it can 'hardly be over-estimated'.[24]

It is both illuminating and sobering that many of the concerns most closely associated with contemporary green politics were already being articulated in the Wilhelmine era. The Heimatschutz activists, and Paul Schultze-Naumburg in particular, can claim much of the credit for this. In the final analysis, however, their efforts serve also to highlight why the German green tradition has proved so problematic.

Notes

1. A suggestion most recently made by Simon Schama in his *Landscape and Memory* (London, 1995).

2. See for example Klaus Bergmann, *Agrarromantik und Großstadtfeindschaft* (Meisenheim am Glan, 1970); Jost Hermand and Richard Hamann, *Stilkunst um 1900* (East Berlin, 1967); Jost Hermand, *Grüne Utopien in Deutschland. Zur Geschichte des ökologischen Bewußtseins* (Frankfurt am Main, 1991); Matthew Jefferies, *Politics and Culture in Wilhelmine Germany: The Case of Industrial Architecture* (Oxford, 1995); Edeltraud Klueting (ed.), *Antimodernismus und Reform. Zur Geschichte der deutschen Heimatbewegung* (Darmstadt, 1991); Stefan Muthesius, 'The origin of the German conservation movement', in Roger Kain (ed.), *Planning for Conservation* (London, 1981); Christian F. Otto, 'Modern environment and historical continuity', *Art Journal*, (1983), 148–57; Rolf Peter Sieferle, *Fortschrittsfeinde? Opposition gegen Technik und Industrie von der Romantik bis zur Gegenwart* (Munich, 1984).

3. The only monograph on Schultze-Naumburg is Norbert Borrmann's *Paul Schultze-Naumburg 1869–1949. Maler-Publizist-Architekt* (Essen, 1989).

4. Schultze-Naumburg's first three major publications, which all appeared before his thirtieth birthday, were entitled *Der Studiengang des modernen Malers* (1896); *Die Technik der Malerei: Ein Handbuch für Künstler und Dilettanten* (1898); and *Häusliche Kunstpflege* (1899).

5. Paul Schultze-Naumburg, 1901, quoted by Borrmann, *Paul Schultze-Naumburg*, 18.

6. See Gerhard Kratzsch, *Kunstwart und Dürerbund: ein Beitrag zur Geschichte der Gebildeten im Zeitalter des Imperialismus* (Göttingen, 1969).

7. Paul Schultze-Naumburg, 1901, quoted by Borrmann, *Paul Schultze-Naumburg*, 25.

8. Lewis Mumford, quoted by Borrmann, *Paul Schultze-Naumburg*, 240 note 889.

9. Borrmann, *Paul Schultze-Naumburg*, 17.

10. Paul Schultze-Naumburg, *Kulturarbeiten*, 7 (*Die Gestaltung der Landschaft durch den Menschen*) (Munich, 1916), 283.

11. Paul Schultze-Naumburg, *Kulturarbeiten*, 4, Städtebau (Munich, 1906), 259.

12. On the organizational history of the Bund Heimatschutz, which claimed up to 100,000 individual and corporate members in 1906, see Otto, 'Modern environment'; Klueting (ed.), *Antimodernismus und Reform*.

13. Ernst Rudorff, 1880, quoted in Hans-Günther Andresen, *Bauen in Backstein. Schleswig-Holsteinische Heimatschutz-Architektur zwischen*

Tradition und Reform (Heide i.H., 1989).

[14.] See Matthew Jefferies, *Politics and Culture*, ch. 2.

[15.] Paul Schultze-Naumburg, 'Heimatschutz. 1. Die Laufenburger Stromschnellen', in *Der Kunstwart*, 18, 1 (1904), 22.

[16.] Paul Schultze-Naumburg, 'Kraftanlagen und Talsperren', *Der Kunstwart*, 19, 15 (1906), 135.

[17.] Schultze-Naumburg, 'Kraftanlagen und Talsperren', 132.

[18.] For further discussion of Heimatschutz legislation, see Matthew Jefferies, *Politics and Culture*, ch. 2.

[19.] Schultze-Naumburg, 'Kraftanlagen und Talsperren', 132.

[20.] See for example Franz-Josef Brüggemeier and Thomas Rommelspacher (eds.), *Besiegte Natur. Geschichte der Umwelt im 19. und 20. Jahrhundert* (Munich, 1989).

[21.] Eugen Gradmann, quoted by Arne Andersen, 'Heimatschutz: die bürgerliche Naturschutzbewegung', in Brüggemeier and Rommelspacher (eds.), *Besiegte Natur*, 152.

[22.] From letter of 15 November 1920 in Krupp Archives, HA Krupp FAH IV E 1096 Paul Schultze-Naumburg 1910–33.

[23.] See Barbara Miller Lane, *Architecture and Politics in Germany 1918–45* (Cambridge, Mass., 1968).

[24.] Julius Posener, *Berlin auf dem Wege zu einer neuen Architektur. Das Zeitalter Wilhelms II* (Munich, 1979), 191.

3 • 'Community Plants the Forests of Justice': Gustav Landauer, Ecosocialism and German Expressionism

Rhys W. Williams

In 1986, in his history of ecological movements in Germany, Ulrich Linse distinguished two trends of green politics: the reformist ecosocialism of the Social Democratic tradition and the radical ecosocialism of a strain of thought which he rightly traces back to the German anarchist tradition, and to Gustav Landauer in particular.[1] Linse concentrates in his account on the importance of Landauer's ideas for the notion (if not the practice) of rural settlements as alternatives to the capitalist industrialization of the late nineteenth and early twentieth centuries. While Linse's interest is focused on the relationship of green thought to anti-militarism and the peace movement, his is the first study to point a direct connection between Gustav Landauer and green consciousness. It will be my contention here that Landauer's influence on the green movement was indeed significant, but that the influence which he exerted was mediated through the work of the creative writers of the Expressionist generation. Landauer's vision helped to shape the assumptions and sensibility of Expressionist writers, not, it must be said, in isolation, but assimilated to other popular contemporary theories, such as the aesthetics of Wilhelm Worringer. The revolution which so many Expressionists proclaimed through their dramas was not a Marxist-Leninist revolution, but a call to spiritual transformation as the prerequisite for social change.[2]

Ernst Toller's experience of the Munich Revolution and the short-lived Räterepublik [Soviet Republic] appear to have been disillusioning in the extreme; he swiftly realized that non-violence and revolutionary solidarity were incompatible values. In his drama *Masse Mensch* [*Masses and Man*] (written in 1919 and first staged in 1921) he conducts a post-mortem on the revolution, and in so doing attempts to bring out the distinction between the socialist ideals which Toller himself subscribed to and the Marxist arguments which he rejected. Indeed, a convincing argument may be (and indeed has been) advanced that the play may be read as a debate between Gustav Landauer[3] (whose revolutionary theories Toller supported) and Marx, between individual and spiritual regeneration on the one hand, and historical determinism on the other. The Woman in the drama represents an ethical, non-violent, anarchist version of socialism, while 'the Nameless One' supplies a Marxist counter-argument. For the Woman, the ideal is not the state, but a community of free individuals ('Gemeinschaft' or 'werkverbundne freie Menschheit');[4] she refuses to define the revolution as merely the seizure of power, advocating instead the quasi-religious spiritual liberation of the masses. In an earlier scene of the play, it is to the Woman that Toller gives the lines:

> Masse soll Gemeinschaft sein.
> Gemeinschaft ist nicht Rache.
> Gemeinschaft zerstört das Fundament des Unrechts.
> Gemeinschaft pflanzt die Wälder der Gerechtigkeit. (GW, II, 95)

> [The masses should be community./ Community is not revenge./ Community destroys the foundations of injustice./ Community plants the forests of justice.]

For Toller, then, the tragedy of the revolution of 1918–19 was that the ideals which Landauer had proclaimed were, at a crucial moment, subordinated to the practical political necessities. Indeed, as the Räterepublik began to disintegrate, with the planned communist putsch on 10 April 1919, Landauer wrote to the Action Committee, distancing himself from their activities, as Toller himself records in his autobiography *I was a German*:

> By the struggle to bring about conditions which will permit each person to partake of the products of the earth and of culture, I understand something very different from you. The advent of socialism will unleash at once all the creative impulses; in your work, however, I perceive that in the economic and cultural sphere you have no expertise. Far be it from me to interfere in the slightest with the onerous defensive action in which you are engaged. But I bitterly regret that what is now being proposed bears only the slightest similarity to my aim: that of warmth and uplift, of culture and rebirth. (GW, IV, 150–1)

Whatever his own reservations, Landauer's opponents on the right were unconcerned with such niceties; when the Räterepublik collapsed, he was to fall victim in the most brutal way to the revenge of the counter-revolutionaries.

The quotation which I have adopted for my title not only reflects the importance of Landauer's socialist ideas for an understanding of Toller's dramas, but also contains a recurring image in German Expressionist writing. Humane values – *Gerechtigkeit* [justice], *Gemeinschaft* [community], *Geist* [barely adequately rendered in English as 'spirit'] or any tautological combination of these terms – are frequently presented through nature imagery, in particular through the topoi of *Wald* [forest] and *Garten* [garden], by images of nature in its primal or pre-industrial state. What I intend to demonstrate in this chapter, through an examination of Landauer's work in particular, but also by locating Landauer in the context of Wilhelm Worringer's aesthetics, is that this cluster of nature imagery was imbued with a more specific political and aesthetic significance than has been hitherto appreciated. The forests and the gardens of German Expressionism are, by implication at least, socio-political forests and gardens. While Landauer's ideas cannot, of course, be equated unambiguously with the ecological concerns of the green movement today, they operate as prototypes of green thinking and may helpfully be viewed as such. The ambivalence of the progressive and the conservative, which is an aspect of green ideology, is similarly a feature of German Expressionism, a feature, moreover, which has caused critics of Expressionism some difficulties.

Gustav Landauer, described by Peter Marshall in his monumental *Demanding the Impossible: A History of Anarchism* (1992) as 'the most important anarchist thinker in Germany after Max Stirner',[5] was born in Karlsruhe in 1870. He joined the SPD, but was expelled in 1891. From 1892 he edited the Berlin anarchist paper *Der Sozialist [The Socialist]*, to which he added the subtitle *Organ für Anarchismus-Sozialismus [Organ for Anarcho-socialism]*. Although he uses the word 'socialism' to describe his anarchistic principles, he consistently inveighs against Marxism in favour of a libertarian and individualist brand of socialism: Writing in 1911, he defines his anarchism thus:

Anarchy is the term for the liberation of man from the idols of the state and of capital; socialism is the term for the true and genuine connection between people, genuine because it springs from the individual spirit, because it blossoms as a living idea, one and the same, in all men, because it exists as a free bond between men.[6]

The individualism, the centrality of 'spirit', what is elsewhere called 'verbindender Geist' [unifying spirit], and, moreover, the organic image ('blossoms') with which the definition closes, all these features recur in the literature of German Expressionism. Landauer's most influential political essays belong to the first decade of the century: his long essay *Die Revolution* [*The Revolution*] appeared in 1907 and his *Aufruf zum Sozialismus* [*Call to Socialism*], given as a lecture in 1908, was published in 1911. Within these two works is contained the essence of that theory of spiritual revolution which Landauer sought to bring about. *The Revolution*, which was published as the thirteenth volume of the series *Die Gesellschaft* [*Society*], edited by Martin Buber (through whom, incidentally, Landauer influenced the Israeli communitarian movement), is concerned, above all, to define *Geist* as the necessary basis of culture in a nation. Landauer's most vivid illustration of a people or nation under the influence of active *Geist* is Christianity, the Christianity of the Middle Ages:

A high level of culture comes into being in situations in which the unity amid the plurality of organizational forms and supra-individual structures is not enforced from outside, but is the product of a spirit inherent in individuals and transcending man's earthly, material interests.[7]

Christianity becomes a 'Geist' which permeates all the social organizations of the medieval period and welds the diverse human functions into a cultural unity. *Geist* itself does not bring about the social organization; however, given the appropriate circumstances, *Geist* will prevail, but only if it can depend on the existence of a set of freely determined mutual relationships, such as were found in the medieval city-state, or the medieval guild. To Landauer's profound enthusiasm for the Middle Ages, I shall return later. Like many influential figures in the early years of the century Landauer sees the medieval world as the high point of western culture, from which there has been a steady decline. The importance of *Geist*, he suggests, diminished with the Renaissance, with its new sense of individualism. Variations on Landauer's arguments were to reappear, incidentally, in the work of Hugo Ball, whose *Zur Kritik der deutschen Intelligenz* [*Critique of German Intellectuals*] (1917) similarly indicts the whole of the western tradition since the Renaissance. Landauer argues that the whole of western history since the Middle Ages has been a long struggle to regain the stability, the hegemony of *Geist*. All subsequent revolutions, the *Bauernkriege* [Peasant Wars], the English Revolution, The Thirty Years War, the American Revolution, the French Revolution, even the Franco-Prussian War, are seen as attempts to bring about the reign of *Geist*. The struggle of *Geist* against dogmatism is the long march of history. The modern political state, Landauer argues in *The Revolution*, arises only when the medieval *Geist* diminishes: it has various forms, despotic absolutism, absolutism of the law and, finally, the absolute power of nationalism. Landauer's anarchistic version of socialism rests on the assumption that the individual can simply refuse his support to the state and its rule will crumble.

With his *Call to Socialism* Landauer attempted to develop a more precise set of articles of faith for the Sozialistischer Bund [Socialist League], which his treatise would directly create. The League envisaged a non-violent social revolution, based on individual commitment to a new notion of social community. Landauer's starting-point is one of cultural pessimism: 'We assert that the peoples of Europe and America have, for a long time, roughly since the discovery of America, been degenerate peoples.'[8] Since the Christian Middle Ages, then, the decline of 'verbindender Geist' [unifying spirit] has been inexorable. Landauer's vision of socialism is vehemently anti-materialist and anti-Marxist. For Marx, he argues, communism will grow out of capitalism, capitalism thus becoming an essential phase in the development of communism. For Landauer, by contrast, capitalism is itself the enemy, 'with its bleak factory system, with its desolation of the countryside, with its masses uniform in their misery, with its economies geared to the world market rather than to the fulfilment of real needs' (*AS*, 47). He wryly notes that Marxism has 'a desperate similarity to military and bureaucratic systems' (*AS*, 50), and seems to thrive in the 'those countries of the army corporal and petty bureaucrat' (ibid.), namely Prussia and Russia. Contemporary readers might regard this as an extraordinarily prescient analysis of the GDR; it certainly prefigures more recent critiques, such as Rudolf Bahro's *The Alternative in Eastern Europe* (German edition 1977, English translation 1978). Landauer is struck by the similarities which exist between capitalism and the economic structures advocated by Marxism; the state capitalism which thus comes into being is necessary only because *Geist* is lacking, 'because justice and love, economic ties and the thriving multitude of small social organisms have disappeared' (*AS*, 56). Here he introduces, for the first time in this essay, an argument in favour of a small-scale sustainable economy. Capitalism, he insists, is worse than slavery, since at least the slave-owner takes care of his investment, while the capitalist is indifferent to the fate of his workers, who become 'cogs in the machine' (*AS*, 91). Turning his attention to economics, Landauer isolates three forms of modern slavery: landowning, the circulation of goods which do not satisfy consumption, and surplus value. The danger of money lies not only in its capital growth, but also in its permanence, its refusal to be consumed. He advocates the adoption of the theory of Silvio Gesell (later to become, for a brief period, economics minister in that ill-fated Räterepublik of which Landauer became, even more briefly, the Education Minister) that a form of money should be invented which loses its value and which must therefore be exchanged for goods as soon as possible thus maximizing its circulation. As regards surplus value, Landauer uses the term 'value' in the sense 'correct value, true value' (*AS*, 123). This value is not the same as the price. The demand which he makes is 'that the price should be equal to the value' (ibid.). Marxists are mistaken, he argues, when they place all the emphasis on work and working conditions to explain the disparity between the value of goods and their price. If they looked further, they would discover that the root of the problem lies in the 'permanence of money, in the way in which its value is retained' (*AS*, 127). Surplus value does not arise at a specific point in the production of goods, but through the circulation of money itself. The solution – and this might strike the modern reader as somewhat idealistic – can come only 'through workers refusing to play their part as capitalist producers' (*AS*, 129); what is needed is 'the rebirth of the people out of the spirit of community' (*AS*, 130).

Having devoted the bulk of his essay to a diatribe against Marxism, Landauer introduces his own ideals in a section prefaced by a lavish tribute to Pierre Joseph Proudhon, who presides over the remainder of the text as a positive counterpart to Marx. The unit which he envisages as the ideal is larger than the family, for 'the family is concerned only with private interests' (*AS*, 132). The aim of Landauer's variant of socialism is to organize society so that everyone works only for himself, with a system of exchange to regulate needs and supplies. What is needed is 'spiritual revolution' (*AS*, 136–7), for *Geist* is dynamic, constantly challenging rigid forms. The treatise ends with a series of rallying cries: 'The earth must be given back to us. The earth is no one's property. Let the earth be without a master; only then will the people be free' (*AS*, 134); 'The struggle of socialism is a struggle for the land; the social question is an agrarian question' (*AS*, 142). The new society will not spring from a general strike, nor from proletarian action, but from the rejection of money as a god. There are hints here of the idealization of the medieval village community, self-sufficient and operating by barter, producing only what the community needs. His aim is to create within capitalist society socialist communities which operate according to this pattern, and he ends his treatise with a call to his audience individually to join such communities, withdraw their support in practical terms from capitalist society and put into practice the idea of a socialism free from the state. What Landauer advocates, with increasing insistence in later essays, is a decentralized, small-scale community, producing only enough to satisfy its needs. Exemplary co-operative settlements on the land and the rejection of urban industrialism are the basic tenets of his socialism: 'The desire to create small groups and communities of justice will bring about socialism, will bring about the beginning of a true society' (*AS*, 99–100).

Perhaps it was Landauer's disillusionment with the German industrial proletariat that prompted his increasing emphasis on agrarian reform. In the 1919 preface to the second edition of *Aufruf zum Sozialismus* (which was reprinted in the fourth edition of 1923) this aspect is highlighted: 'Our revolution can and must distribute parcels of land in large measure; it can and must create a new and renewed peasantry' (p. xiii). His final conviction is 'that, for our salvation and for the sake of acquiring justice and community, we must return to a rural existence and to a new unity of industry, crafts and agriculture' (p. xvi). Landauer was, of course, not without his antecedents: he had translated Kropotkin's *Mutual Aid*, and also extolled British traditions of anarcho-socialism: William Morris's Socialist League in the 1880s embodies a number of features, particularly a Romantic medievalism, which Landauer adopted. As early as 1904, together with his wife, Hedwig

Lachmann, Landauer translated and edited Oscar Wilde's *The Soul of Man under Socialism*, a work which combines socialist and individualistic ideas in a manner calculated to appeal to both Landauer and the Expressionist generation. In that essay Wilde argues that socialism, in freeing man from competitive and acquisitive drives, will permit the individual to fulfil his artistic and spiritual aspirations. Wilde, too, rejects authoritarian socialism in favour of a free and voluntary association of individuals. It was, then, Landauer's fusion of anarcho-socialist ideas with a critique of industrialization, as well as the undoubtedly visionary quality of his style, which appealed to the Expressionists. Of the extent of that appeal, there is ample evidence, as the columns of *Die Aktion* bear witness. The writers of the *Aktion* group could have been expected to approve of Landauer's opposition to the Social Democratic Party, his rejection of bureaucratic centralism, his championing of the organic community, his pacifism and his admiration for figures such as Tolstoy and Walt Whitman.

Of Landauer's importance for Toller's ideas, there can be little doubt. Toller was inspired to write to Landauer in 1917 in terms which suggest that he had taken Landauer's idea of a holistic approach to revolution, his use of the term *Geist*, and his agrarian solution (with its imagery) very much to heart: 'I want to permeate the living, I want to plough it up with love, but also, if I must, I want to overthrow rigid structures for the sake of the spirit.'[9] Toller's enthusiasm for Landauer's social vision emerges powerfully in his drama *Transfiguration* (begun in 1917, completed in 1918, and first performed in 1919). A powerful anti-war drama, *Transfiguration* presents an antithesis between the nation-state and the true community of free individuals. Friedrich's journey is a quest for community; he does not find it in the family, nor by submerging his individuality in nationalism. His ideal of patriotism is shattered by his wartime experiences, and he logically destroys the statue of the goddess of victory which he was engaged in sculpting. Toller, through Friedrich, rejects the institutions of the capitalist state and offers instead a vision of a new community, which will be achieved when men rediscover their common humanity. The revolutionary rallying-cry with which the play ends is pure Landauer: the creative *Geist* has been submerged by capitalist industrialization; only when this inner humanity is released will social relationships be transformed. What Friedrich proclaims is not merely a spiritual revolution, it is highly political; but in an anarcho-socialist, rather than a Marxist sense. The last scene of the play contains an apotheosis of the New Man: Friedrich, addressing the multitude, preaches a Landauer gospel of *Geist*: 'Und ihr könntet doch Menschen sein, wenn ihr den Glauben an euch und den Menschen hättet, wenn ihr Erfüllte wäret im Geist' ['And yet you could be true human beings, if you but believed in yourselves and in man, if you were but fulfilled in the spirit'] (*GW*, II, 60). The final cry of 'Revolution! Revolution!' (*GW*, II, 61) is clearly not the revolution envisaged by Marx; it involves a quasi-religious spiritual awakening, a rejection of both the war machine, and the related machinery of capitalism. The crucifixion of man which symbolizes his suffering under capitalism and chauvinism is followed by the resurrection of the spirit.

Even more than *Transfiguration*, Toller's *Masses and Man* confronts the Landauer/Marx debate. The anarchist arguments are rehearsed from the first scene onwards: the Woman rejects the authority of the state and that of her husband:

> Dein Staat führt Krieg,
> Dein Staat verrät das Volk!
> Dein Staat ausbeutet, drückt, bedrückt, entrechtet Volk.

> ['Your state conducts war,/ Your state betrays the people!/ Your state exploits, forces and oppresses the people, denying them their rights.'] (*GW*, II, 72)

Subsequent scenes confront us with the mechanics of the financial world, and then the exploitation of the factory: echoes of Landauer abound in the complaint of the factory girls:

> Wir aber siechen, von Scholle entwurzelt,
> die freudlosen Städte zerbrechen unsre Kraft.
> Wir wollen Erde!
> Allen die Erde.

> ['But we are wasting away, uprooted from our native soil,/ the joyless city shatters our strength./ We want the earth!/ To all the earth.'] (*GW*, II, 82)

The revolutionary transformation which the drama presents in the final scene embodies a shift in values. The Woman goes to her death, yet her exemplary self-sacrifice transforms the consciousness of two female prisoners who enter her cell to steal her belongings. Once more the revolution in question is a spiritual one, shot through with religious symbolism.

Ernst Toller was not alone in his attempt to give literary expression to Landauer's revolutionary ideals. Georg Kaiser, too, owed a large debt of gratitude to Landauer, both intellectually and personally. By 1916, when Kaiser and Landauer began a correspondence, Landauer had already written a short piece on Kaiser's *The Burghers of Calais* in his periodical *Der Sozialist* [*The Socialist*], for it was to thank Landauer for an offprint of this essay that Kaiser first wrote to him on 1 March 1916.[10] Indeed, it was thanks to Landauer's essay (published on 6 February 1916) that Kaiser's play was accepted for performance; it was Kaiser's first great dramatic success. Over the next three years Kaiser and Landauer corresponded regularly, Kaiser frequently voicing his admiration for Landauer's work. In a letter to Otto Liebscher on 17 May 1919, Kaiser expresses his profound grief at Landauer's murder: 'Gustav Landauer is dead: in his head they have shot heaven in two. I feel that I have been robbed of everything; the previous catastrophes that life has hurled against me are as nothing compared to this crippling of my very being.'[11] The relationship between Kaiser and Landauer was, for the last three years of Landauer's life, one of mutual respect and admiration. On the impact of Landauer's ideas on Kaiser's *From Morn to Midnight* I shall touch here only briefly.[12] The whole play offers a set of variations on the theme of value, on the contrast between what Landauer calls 'true value' and the value placed on things in a capitalist economy. After embezzling the bank's money, the *Kassierer* [Cashier] sets out in search of a new value scheme to replace the one that he has lost, but, since his unit of exchange remains money, his quest will be doomed to failure. He experiences the excitement of organized sport in the six-day cycle race, but is horrified to discover that the spectators are distracted from their excitement by the arrival of the Kaiser. The 'unifying spirit' ('verbindender Geist') which seemed to be present in the stadium is superseded by nationalism, a mere surrogate in Landauer's eyes. The *Kassierer*'s final insight into Landauer's thesis, that the problem of modern society lies in the nature of money itself, prompts his passionate denunciation: 'Mit keinem

Geld aus allen Bankkassen der Welt kann man sich irgendwas von Wert kaufen. Man kauft immer weniger als man bezahlt' [In all the bank accounts of the world there is no money with which one can buy anything of value. What you buy is always worth less than you pay].[13] One interesting detail of the text is that the *Kassierer*'s ideals are presented in terms of an unknown Lukas Cranach painting of the Fall. The art historian son of the Italian Lady responsible for jolting the *Kassierer* out of his submission to capitalism, discovers the Cranach, thus prompting his mother's visit to the bank to raise the money for the purchase and setting in train the events that follow. When the son describes the painting of Adam and Eve in the garden as 'der wirkliche Sündenfall' [the real fall of Man], he little realizes how accurate he is. The *Kassierer*, throwing off the shackles of capitalist society, reverts to a pre-civilized state of innocence, seeing himself as Adam, even as he is to be transformed into the crucified Christ at the end of the play. Spotting the painting in the Lady's hotel room, the *Kassierer* manages, through a grotesque misunderstanding, to identify himself and the Lady as Adam and Eve. Perhaps it is the Lady's injudicious remark that prompts the confusion:

DAME: Dies Bild steht in enger Beziehung zu meinem Besuch auf der Bank.
KASSIERER: Sie?
DAME: Entdecken Sie Ähnlichkeiten?
KASSIERER (*lächelnd*): Am Handgelenk!
DAME: Sind Sie Kenner?
KASSIERER: Ich wünsche – mehr kennenzulernen.
DAME: Interessieren Sie diese Bilder?
KASSIERER: Ich bin im Bilde!

[LADY: This painting is closely connected with my visit to the bank.
CASHIER: You?
LADY: Do you spot any similarities?
CASHIER (*smiling*): The wrist!
LADY: Are you a connoisseur?
CASHIER: I wouldn't mind getting to know more.
LADY: Do these paintings interest you?
CASHIER: I'm in the picture.] (*W*, I, 477)

Later, in the last scene of the play, shortly before his betrayal, when he believes that he is to experience with the Salvation Army girl the kind of fulfilment of which he has dreamed, the biblical analogy reappears: 'Mädchen und Mann. Uralte Gärten aufgeschlossen' ['Maid and man together. Age-old gardens opened up'] (*W*, I, 516). The garden, whether it be Eden or Gethsemane, is the setting for that experience of freedom from the constraints of an industrial society, a stripping away of the accretions of an inimical, indeed, threatening society.

The notion of a flight from an over-industrialized and mechanistic European society is connected with Landauer's analysis of its ills, and with his vision of transformation. The fact that Lukas Cranach is the immediate inspiration here points to another major influence on Expressionist writing which deserves mention, namely the asethetics of Wilhelm Worringer. Worringer's doctoral dissertation, entitled *Abstraction and Empathy: A Contribution to the Psychology of Style*, first published in 1907[14] as a thesis, was, somewhat unusually, recommended to, and reviewed favourably by, the writer Paul Ernst, whose article in *Kunst und Künstler* [*Art and Artists*] swiftly prompted a publication in book form the following year. It is a peculiar feature of Worringer reception that his first publication not only established his reputation as a significant art historian, but has seldom been out of print since 1908. The 1976 edition is currently in its fourteenth reprint. That a doctoral thesis should have had such a profound impact, not only for art historians and theorists, but also for generations of creative writers and intellectuals, is almost unprecedented.

Abstraction and Empathy takes as its starting-point Theodor Lipps's theory of empathy, the notion that the work of art maximizes our capacity for empathy, that beauty derives from our sense of being able to identify with an object. While conceding that a mimetic urge exists in man, Worringer denies any necessary connection between *mimesis* and art: if Egyptian art was highly stylized, he argues, this was not because its artists were incompetent and failed to reproduce external reality accurately, but because Egyptian art answered a radically different psychological need. In mimetic works, he argues, we derive satisfaction from an 'objectified delight in the self';[15] the aim of the artist is to maximize our capacity for empathizing with the work. This kind of art springs from a confidence in the world as it is, a satisfaction in its forms, which is embodied in Classical and Renaissance art. By contrast, the urge to abstraction, exemplified variously by Egyptian, Byzantine, Gothic or primitive art, articulates a wholly different response to the universe: it expresses man's insecurity and seeks to answer transcendental or spiritual needs. In certain historical periods, Worringer argues, man is confidently assertive and finds satisfaction in 'objectified delight in the self' and can abandon himself in contemplation of the external world; but, in periods of anxiety and uncertainty, man seeks to abstract objects from their contingency, transforming them into permanent, absolute, transcendental forms.

It is not difficult to appreciate the fascination which Worringer's theory held for the writers and intellectuals of the Expressionist generation in Germany. He supplied a theoretical justification, and an aesthetic category, for the revival of interest in early German art and for the simultaneous discovery of primitivism. His theory also had a sociological dimension. The fear and alienation which 'primitive man' had experienced in the face of a hostile universe had, he argued, prompted an urge to create fixed, abstract, and geometric forms; while Renaissance individualism had later weakened the capacity for abstraction, the modern experience of industrialization, the notion that individual identity was threatened by a hostile mass society, had rekindled the need for abstract forms to counteract a sense of what Marx (and Landauer) had defined as alienation.

Worringer's second major publication, *Form in Gothic* (1911; English translation 1923), developed the closing section of his earlier thesis into a fully-fledged account of Gothic art and architecture. Once again, Worringer operates with a definition of the Gothic which is all-embracing: 'Gothic', for him, is all the art of the western world which was not shaped by classical Mediterranean culture, all art, in short, which was not a direct reproduction of nature. Starting with the latent Gothic in early northern ornament, with its repetition, its geometric pattern, its flowing line, Worringer moves on to examine cathedral statuary, and finally Gothic architecture, with its 'will to expression, striving towards the transcendental'.[16] Where Worringer attempts to locate his theoretical abstractions in specific historical contexts, the modern reader may be uncomfortably reminded of the misuse of such theories under National Socialism. Statements like 'the land of the pure Gothic is the Germanic north',[17] a formulation which actually reflects Worringer's fondness for employing antithetical terms, like north/south, Gothic/Renaissance, abstraction/empathy, have acquired unfortunate ideological connotations. Perhaps this factor, too, helps to explain why Worringer's work has, like Landauer's, suffered from comparative critical neglect. My argument here is that Landauer's diagnosis of the ills of modern society and his advocacy of a spiritual revolution meshes with Worringer's aesthetic of a spiritual and transcendent art. Thanks to the temporal (and intellectual) proximity of the work of Landauer and Worringer the rediscovery of both primitivism and Early German art is given a socio-political dimension.

My purpose is not merely to assert Landauer's influence, but to develop another set of arguments: namely, that Landauer's concept of revolution offers some antecedents for green thought, and, moreover, that his writings helped to shape the metaphoric language of Expressionism, exerting indirectly an impact even on writers who were temperamentally or politically unsympathetic to his ideas. If industrialization is diagnosed as the problem of modern society, clearly particular interest is shown in the pre-industrial, whether that be the medieval world, or, more radically, the South Sea island primitivism of Sternheim in *Tabula rasa* or Benn's evocation of 'Regression'. The imagery and topoi of Expressionist writing fall into two contrasting clusters: on the one hand the negative values connoted by machines, money, militarism, depersonalization, industry and pollution; on the other, the positive spirituality suggested by untamed nature (forests), tamed nature (gardens), medieval settings, religion, and both sacred and profane love. On the one hand, the mechanistic and the rational, on the other the organic and the intuitive.[18] These clusters operate associatively, the presence of any one element often suggesting the presence of another from the same cluster. These antinomies operate quite explicitly in the works of Toller and Kaiser, Sternheim and Benn; but they are also present, at least by implication, in the work of writers whose political values were far less radical. The argument may be tested out by examining a poem by Georg Heym:

Berlin I

Beteerte Fässer rollten von den Schwellen
Der dunklen Speicher auf die hohen Kähne.
Die Schlepper zogen an. Des Rauches Mähne
Hing rußig nieder auf die öligen Wellen.

Zwei Dampfer kamen mit Musikkappellen.
Den Schornstein kappten sie am Brückenbogen.
Rauch, Ruß, Gestank lag auf den schmutzigen Wogen
Der Gerbereien mit den braunen Fellen.

In allen Brücken, drunter uns die Zille
Hindurchgebracht, ertönten die Signale
Gleichwie in Trommeln wachsend in der Stille.

Wir ließen los und trieben im Kanale
An Gärten langsam hin. In dem Idylle
Sahn wir der Riesenschlote Nachtfanale.[19]

[From the dark openings of the warehouses rolled/ caulked barrels on to the tall barges below./ Tugs took the strain. The manes of smoke/ hung sootily down over the oily waves./ Two steamers with dance bands came by,/ dipping their funnels to pass beneath the bridges./ Smoke, soot and stench lay on the filthy streams/ from tanneries with their brown hides./ Claxons sounded at every bridge/ as the barge passed beneath us,/ like drum rolls growing in the silence./ Casting off, we drifted slowly along/ past gardens. In this idyll/ we saw the flare of giant chimneys against the night sky.]

Written in April 1910 and first published in 1911 in the collection *Der ewige Tag* [*The Eternal Day*], the poem is not Expressionistic in the accepted sense; it contains no violent daemonic imagery of the city, as in 'Der Gott der Stadt' ['The God of the city'] and both the sonnet form and the impressionistic quality of the observation hint at the influence of George. Nevertheless, the poem displays a number of features which make it more typical of an Expressionist sensibility.[20] The opening two quatrains present us with a Berlin cityscape of hectic industrial activity: the sights, the sounds and the smells of industry. The rapid industrial growth of Berlin since 1871 into the capital of the fastest-growing economy in Europe has produced a squalor and pollution which for many presented the unacceptable face of capitalism. What is interesting here is that the first eight lines of the poem contain no references to people. Industrial activity seems to go on without human agency: the barrels roll, the tugs take the strain, the tanneries (surely the filthiest of all industrial processes) release their effluent into the canal. Nor are the leisure activities of the Berlin proletariat forgotten: amid the stench and squalor two steamers go by, with dance bands, playing for their public, again as if without human agency. It is the steamers, rather than any human beings, which are imbued with volition, as they dip their funnels to pass beneath the low bridges. The leisure of the Berlin proletariat is played out amid the very factories in which working life is spent. Leisure is organized as an adjunct to the productive process. Not until the sestet do human beings make a belated appearance, and then only indirectly. Only in the last three lines are the perceiving subjects mentioned, the 'we' through whose eyes the scene has been registered: the poet and a companion, drifting along the canal, past the gardens, the idyllic setting which manages (just) to exist within the overwhelmingly oppressive industrial landscape. The 'we' suggests the possibility of human contact, but only outside the industrial landscape, remote from the anonymous proletariat whose noisy revelry is an appropriate accompaniment to the industrial process. Moreover, the 'we' are depicted as drifting without volition, without goal-directed purpose. The artistic sensibility of the (presumably bourgeois) poet is not part of the industrial scene; he is merely an observer, powerless to affect the industrial process, capable only of momentary escape to a garden idyll, yet aware that this escape is but temporary and incomplete. The backcloth of the factory chimneys ensures that the socio-economic realities remain overwhelming. Further distance between the observers and the cityscape is engendered by the imperfect tense employed throughout. This experience is recollected at a spatial and temporal distance, and, for the moment, successfully confined within the sonnet structure. Only in Heym's later poetry is the 'God of the cities' to emerge in his full destructive and daemonic might. For the moment, the idyllic garden, the vantage point of the poetic imagination, is still a possible, if embattled, refuge. Withdrawal from industrial society can alone bring spiritual and aesthetic renewal.

It is my contention that Gustav Landauer contributed in large measure to the critiques of capitalist society inherent in the Expressionist movement. If his contribution has been neglected, that is largely due to the polarization of German politics in the late 1920s and early 1930s.[21] The Marxist critique eclipsed the many anarcho-socialist variants which so influenced Expressionism, and in so doing rendered Expressionism itself problematic for Marxists, as Lukács's famous assault on the movement illustrates. It is striking that contemporary German theorists like Rudolf Bahro have moved similarly from a critique of Marxism to a green position. Bahro's eco-fundamentalism operates with a concept of 'withdrawal and renewal' which is strikingly similar to Landauer's; for Bahro, too, the establishment of small-scale co-operatives, or 'liberated zones', provides refuge from the depredations of industrialization. As Robyn Eckersley has insisted, 'Bahro is at pains to point out that the challenge of ecological degradation is primarily a cultural and spiritual one and only secondarily an economic one. Accordingly we must direct our attention to cultural and spiritual renewal rather than structural or economic reform.'[22] Clearly, it would be absurd to attempt to derive contemporary German ideas on eco-communalism exclusively from Gustav Landauer. What is beyond doubt, however, is that Landauer exerted a powerful fascination for the writers of the Expressionist generation, helping to shape both their cultural pessimism and their visions of spiritual renewal. The polarization of German politics, exacerbated by the division of Germany, ensured the continued neglect of Landauer's contribution. Unification, combined with the renewed urgency of the ecological debate, has witnessed the partial rediscovery and rehabilitation of a figure who was of seminal importance in the first two decades of the twentieth century.

Notes

1. Ulrich Linse, *Ökopax und Anarchie. Eine Geschichte der ökologischen Bewegungen in Deutschland* (Munich, 1986).
2. See also Axel Goodbody on catastrophism and the Expressionist generation (pp. 164–5 of this volume).
3. Useful studies of Landauer include Wolf Kalz, *Gustav Landauer: Kultursozialist und Anarchist* (Meisenheim am Glan, 1967); Charles B. Maurer, *Call to Revolution: The Mystical Anarchism of Gustav Landauer* (Detroit, 1971); Eugene Lunn, *Prophet of Community: The Romantic Socialism of Gustav Landauer* (Berkeley and Los Angeles, 1973); and Peter Marshall, *Demanding the Impossible: A History of Anarchism* (London, 1992).
4. Ernst Toller, *Gesammelte Werke*, ed. John M. Spalek and Wolfgang Frühwald, six vols. (Munich, 1978), II, 110. Subsequent quotations from Toller are taken from this edition; references are indicated by the abbreviation *GW* and appear in parentheses after quotations.
5. Marshall, *Demanding the Impossible*, 410.

6. 'Individualismus', reprinted in Gustav Landauer, *Auch die Vergangenheit ist Zukunft. Essays zum Anarchismus*, ed. Siegbert Wolf (Frankfurt am Main, 1989), 138–44, here 144.

7. Gustav Landauer, *Die Revolution*, 2nd edn. (Frankfurt am Main, 1919), 40. Published as Vol.13 of *Die Gesellschaft. Sammlung sozialpsychologischer Monographien* [*Society: A Collection of Socio-psychological Monographs*], ed. Martin Buber.

8. Gustav Landauer, *Aufruf zum Sozialismus*, 4th edn. (Cologne, 1923); reprinted by the Verlag Büchse der Pandora (Wetzlar, 1978). All subsequent quotations refer to this edition; page references are indicated by the abbreviation *AS* and are given in parentheses after quotations. For a useful introduction to the text, see G.L., *Aufruf zum Sozialismus*, ed. H. Heydorn (Frankfurt am Main, 1967).

9. 'Brief an Gustav Landauer', in Toller, *Gesammelte Werke*, I, 35.

10. Georg Kaiser, *Briefe*, ed. Gesa M. Valk, with an introduction by Walter Huder (Frankfurt am Main, Berlin and Vienna, 1980), 107–8.

11. Kaiser, *Briefe*, 168.

12. For a detailed exposition of Landauer's influence on the play, see my article 'Culture and anarchy in Georg Kaiser's "Von morgens bis mitternachts"', *Modern Language Review*, 83 (1988), 364–74.

13. Georg Kaiser, *Werke*, ed. Walter Huder, six vols. (Frankfurt am Main, Berlin and Vienna, 1971–2), I, 515. All subsequent quotations from Kaiser's work refer to this edition; page references are indicated by the abbreviation *W* followed by the volume number and are given in parentheses after quotations.

14. The English translation by Michael Bullock appeared in 1953.

15. Wilhelm Worringer, *Abstraktion und Einfühlung* (Munich, 1987), 47.

16. Wilhelm Worringer, *Formprobleme der Gotik* (Munich, 1920), 71.

17. Ibid., 97.

18. It is worth noting that these same clusters of images employed antithetically find their way into German writing in the 1950s and 1960s. Heinrich Böll's *Das Brot der frühen Jahre*, in particular, operates with a virtually identical set of contrasting terms: on the one hand the negative values of the 'economic miracle', greed, mechanization, money (and the colour red); and on the other, positive spiritual values, bread, kindness, wild landscapes (and the colour green). In Böll's case, social values are at variance with humanitarian or existential ones, so much so that moral actions are, almost by definition, antisocial actions. It is this feature of Böll's writing which prompted J. P. Stern to write of Böll's 'gentle anarchy'.

19. Georg Heym, *Dichtungen und Schriften*, ed. Karl Ludwig Schneider, 4 vols. (Hamburg and Munich, 1964), I, 58.

20. Patrick Bridgwater, *Poet of Expressionist Berlin: The Life and Works of Georg Heym* (London, 1991), concludes: 'concentrated and powerful

though it is, the poem cannot be called "expressionistic" in any sustainable sense' (205). My analysis is intended to suggest that there is more Expressionism in the poem than meets the eye. I have supplied my own translation of the sonnet; it differs from the prose version offered by Bridgwater in a number of respects: Bridgwater translates 'den Schornstein kappten sie' as 'their funnels hit the arch of the bridge', while I remain convinced that the ships' funnels are hinged and are simply swung down to enable them to pass beneath the bridges. Bridgwater also locates the 'wir' on the barge in 'drunter uns die Zille hindurchgebracht', while I locate them on the bridge; only later in the poem do they take to the water ('wir ließen los') and then on an unpowered vessel.

21. Raymond H. Dominick III, *The Environmental Movement in Germany: Prophets and Pioneers 1871–1971* (Bloomington and Indianapolis, 1992), unaccountably makes no reference to Landauer's contribution to environmental thought.

22. See Robyn Eckersley, *Environmentalism and Political Theory: Towards an Ecocentric Approach* (London, 1992), 164. Eckersley points out that Bahro's position has shifted from an ecosocialist to an eco-communalist one.

4 • The Cultural Impact of Green Thought under National Socialism: The Case of Lothar Schreyer

Brian Keith-Smith

> Nature mirrors inner human life, but it is not its mirror image. We can indeed mirror ourselves in nature, but then we see our image in the mirror and not that of nature.[1]

The role of proto-green thought in Nazi ideology has been examined by several historians, usually with special reference to political and social developments in the 1930s and early 1940s and to the background of some late nineteenth-century thought. In particular, Daniel Gasman has highlighted the importance of Ernst Haeckel and the influence of monism on Adolf Hitler, especially through the social Darwinism of Haeckel's disciple and biographer Wilhelm Bölsche, and of Haeckel's admirer the Norwegian explorer Fridtjof Nansen with his glorification of nature and emphasis on distinctive elements in pre-Christian northern European culture.[2] Gasman also speculates that Hitler's interest in the book *Urwelt, Sage und Menschheit* [*Primeval World, Saga and Humanity*], probably written by the nature mystic Edgar Dacqué, along with his general approval of Dacqué's ideas on the links between evolution, biology and human culture, point to sources out of which developed Hitler's belief in the preservation and maintenance of Germanic biological superiority. Hitler's eugenics, anti-Semitism and conception of history developed monist views to one extreme when he argued that 'Western Civilization had obscured the true relationship between man and nature.'[3] The extreme lay in Hitler's attacks on the development of the Christian church quoting the Papacy's intolerant persecutions and Christianity's 'rebellion against natural law, a protest against nature'.[4] Gasman foregrounds Hitler's belief in observing nature in order to destroy the world of superstition, and claims that 'Hitler ended his life as an evolutionary Monist in the deepest sense.'[5]

Anna Bramwell has highlighted the development of 'green' Nazis in her book *Blood and Soil: Richard Walther Darré and Hitler's 'Green Party'* (Abbotsbrook, 1985), and pointed out how Darré used economic as well as moral arguments to support a locally based economy. By living closer to nature, Darré believed, man could guard against the more harmful effects of over-civilization. Science should be used to improve society and make man less dependent. His programme as Minister of Agriculture in the Third Reich of a 'biological-dynamic agriculture' amounted to a rejection of industrial capitalism and a utopian hope that in a victorious post-war world Germany could explore the practical implications of his theories. In the event, Darré's limited experiments collapsed even before the demise of Nazi Germany, when he was elbowed aside by rivals in the early war years.

Whereas Haeckel's anti-Christian writings and Darré's political experiments give some evidence of proto-green ideology within the Third Reich, Lothar Schreyer's writings during the late 1920s and the early 1930s provide one example of the often confused situation in which many Germans found themselves at the time. Usually presented as an Expressionist, a *Meister der Form* [master of form] at the Weimar Bauhaus and, later in his life, as a committed Christian artist and writer, he was also a nature fanatic with some eccentric views on ecology. Early interpretations of nature as an artistic overall view were supported by his lifelong interest in various forms of mysticism, not least in his works on Jakob Böhme.[6] His unpublished notebooks and letters frequently refer to his love of the landscape and in particular the sea; regular holidays on the island of Sylt and his emphasis on escape from the civilization of the modern city, in his case Dresden, Berlin and Hamburg, provided not only physical but also spiritual regeneration. His views were adapted for *Deutsche Landschaft. Eine Naturbetrachtung* [*German Landscape: Contemplating Nature*], and further explored in other works published at the time.[7] Taken individually, these works relate one aspect of nature, art or civilization to an all-embracing spiritual sphere. Taken together, the holistic character of Schreyer's thought becomes apparent. Active involvement with Roman Catholicism resolved for him the apparent confusion of the late 1920s and early 1930s, as becomes clear not only from unpublished poems, but also from two major typescripts of the 1950s. *Über die Grenzen der Natur* [*Beyond the Bounds of Nature*], with collected notes from writers, philosophers and artists and his own comments on various aspects of nature, and *Heilige Weltschau* [*Divine World Revelation*] prove that his more private thoughts were directed towards commitment as a Christian and Catholic. Whatever reservations he may have shown in the 1920s and early 1930s about church history, Schreyer's personal faith later flourished and was given prolific expression in his analyses of religious art and publication of catechetical books and artistic and literary works on various saints and devotional subjects.[8]

The evidence for his active interest in nature comes mainly from the volumes he wrote for the Hanseatische Verlagsanstalt publishing house in Hamburg in the early 1930s, where he was the reader and main editor for culture and literature from 1928 to 1932, but also from some of his several dozen poems, many as yet unpublished, where we can find ample justification for him as a nature poet. This can be supported by early unpublished prose works such as *Der Märchenwald* [*The Fairytale Forest*] and *Die weiße Lilie* [*The White Lily*], and several short essays, especially from the 1940s. Themes from nature also frequently form the basis of his interpretations of art, most strikingly in his introductory comments to *Hagen malt in Hiltfeld* [*Hagen Painting in Hiltfeld*], a catalogue appearing in 1963.

Already in his childhood he tells us in his diary that during his summer holidays in the 'Waldaltelier' [forest studio] of his artist father on the heath at Niederlausitz he learned to understand nature in an artistic context. The natural landscape meant above all for Lothar Schreyer *the* major factor in human experience that could give pattern and personal satisfaction to every individual human life, a process that he at first related to art. His interest in the landscape led him later to develop an intensely religious inner life, to immerse himself in the study of mystic texts and to write long catechetical and private works of worship. Indeed, his claim in retrospect that at six years of age 'I felt and sensed indeed a spiritual reality behind everything, which of course was incomprehensible yet could be given expression'[9] shows how much he strove to see all natural phenomena and works of art as proofs of hidden meaning.

Unfortunately, this eventually devotional regard for nature was clouded over for some years when he toyed with the blandishments of a proto-fascist nationalism, which is strongly expressed on a few pages of the volume *Deutsche Landschaft*. We can interpret this as a flirtation with extreme right-wing forces that included friendships in the late 1920s: with Wilhelm Stapel, for whom he wrote several articles to be published in Stapel's periodical *Deutsches Volkstum,* with the German nationalists Hans Haffenrichter and Adolf Behne, and with members of the Berlin Groß-Loge Fraternitas Saturni Orient. Schreyer was at the crossroads. In 1928 his old friend Theodor Bogler, who had taught pottery at the Weimar Bauhaus, became a monk at the Benedictine abbey of Maria Laach. At the same time a young Roman Catholic Austrian, Maximilian Tischler, became Schreyer's secretary at the Hanseatische Verlag. In 1932 he visited Maria Laach for the first time, converted to Roman Catholicism in 1933 and eventually became a lay novice of the abbey. Perhaps the most striking sentence that tells us much about his inner struggle at this time came in *Deutsche Landschaft:* 'Consideration of nature purifies the human soul' (*DL*, 37); indeed its emphasis on purification could stand as a subtitle to his essays on art, nature and religion of 1928–9, which were crucial years for him. The implications behind the word purification are the key to Schreyer's struggle. It could, for instance, imply élitism, as in a short essay 'Pferde' ['Horses'], published in 1929 that emphasizes just how close he then came to confusing freedom and bondage, a form of nature worship and praise of the élitist concept of Nordic man.[10] The essay opens as follows: 'Horses are the sacred beasts of the old Germans. They are the beasts of prophecy in the sacred grove. They carry the leader of the peoples into battle. They are sacrificed on the graves of the heroes.'[11] Later we read:

> The natural fate of Nordic man is enfigured in the horse. He knows from his own experience the pale horse of death from the apocalypse. He also knows of the white horse of revelation, on which the hero of faith is said to have appeared in the past.
>
> And even today's man senses his bond with nature whenever he looks at the horse. Does he feel his companion from time immemorial, from whom the man of the present has separated himself?[12]

It is only with this interweaving of religious, cultural and, in some essays, blatantly racist comments in mind that we can understand the strange mixture of statements in *Deutsche Landschaft.* An essay such as 'Ausblick auf eine heroische Kunst' ['Outlook on a heroic art'] (1934) is a reminder of just how far a writer like Schreyer could go, even though he had converted to Roman Catholicism deliberately and symbolically on the day that Adolf Hitler came to power.[13]

Here, he expressly claims the need for the artist to serve the state, for 'The art of our state will pronounce the spiritual reality of our race, whose body is the people.'[14] That art he sees based on *Blut und Boden* [blood and soil], which celebrates the heroic deeds of the German present. These are presented in technical artistic terms as the struggle to produce the ideal expression of Nordic man by emphasizing line rather than surface. In his *Die bildende Kunst der Deutschen* [*The Art of the Germans*] (1931) Schreyer insists that: 'German man lives in the darkness and strives for the light. From time immemorial he wanders through the forests worshipping the spirits of light.'[15] As for today's German: 'We still go on wandering through the forest of our soul and feel ourselves bound to the woods of our homeland.'[16]

The archetypal art for Nordic man, reflecting this struggle between dark and light towards purification, he identifies as black and white woodcuts, a medium, incidentally, which he celebrated in his ten-volume series entitled *Meister der Graphik* [*Masters of Graphic Art*] (l948-9). Indeed, 'Every work of art is a mini-image of the world order that we aspire to, of the re-establishment of the order of creation lost through our indiscipline.'[17] Here, in the essay 'Ausblick auf eine heroische Kunst', we find one root of Schreyer's understanding of nature – that it is not only the proof of hidden meaning but also a challenge in that it needs to be recreated, reordered by man who has been purified in community. Here purification implies for him obedience and discipline. He goes further in his definition of art when he states: 'An art that does not serve the order of the people and the nation is an act of immorality, an atrocity and will not be allowed to be honoured with the honourable title "art"' (p.217). Ideal German art is the symbol of a people's community (*Volksgemeinschaft*). It is this concept that links for Schreyer the purification of man and his reunification with nature. Unfortunately, he ends this same essay by praising the symbol of the *Deutsche Arbeitsfront* [German Workers' Front], the swastika, as *the* sign of purification. Schreyer's way had to go through this symbol before he could reach his later commitment to and frequent representation in his art in many different variations of the Christian cross.

In an earlier paper on 'The concept of *Gemeinschaft* in the works of Franz Werfel and Lothar Schreyer'[18] I examined briefly Schreyer's strange work *Verantwortlich* [*Responsible*][19] in which he declares *Gemeinschaft* [community] to be *the* saving principle for a mankind that has forgotten its real identity. *Gemeinschaft* implies for him (as for Werfel) a change in the individual and only then indirectly a change in mankind in general:

> Community does not arise from the organization of the masses, but it develops out of the growth of the individual ... Community life is not a homogeneous life. It is a living towards one another of discrete individuals ... We must not live close to one another. We cannot live with one another. We want to live towards one another.[20]

Such a statement lies at the heart of Schreyer's beliefs as a theatre director, as a Christian, as an art historian and as a lover of nature.

Most of the volume *Deutsche Landschaft* summarizes and praises the features of the German landscape, its mountains, woods, rivers, heaths and the sea and sky against which it is set. The book however leads towards a section entitled 'Supernatural Homeland' largely consisting of a long quotation from the German mystic Heinrich Seuse, and to a meditation on the graces accorded to the Christian soul in the words of Mechthild von Magdeburg, which inspires Schreyer to formulate the definitive statement:

> The German landscape together with the German nation and its people, led and living in such order, are the Reich. It is a Reich of the earth, a Christian Reich of the German nation. Such a Reich pronounces the heavenly Reich, for in such a Reich hope, love and faith are the forces that with grace have taken hold of human life, which can no longer withstand them. (*DL*, 279)

The first part of the book, however, shows Schreyer contrasting *Naturland* with *Kulturland* with the claim that 'Only those peoples whose own forces, those of culture, are stronger than civilization, can survive a particular epoch' (*DL*, 12). In a section that almost mirrors some of Darré's ideas, he contends that modern technology leads man away from the *Volksgemeinschaft* which owed its development to the coexistence of the *Heimat* and the soul of the German people. This leads on to a section entitled 'Wandering People' where we read:

> Man wandered from nature into the city, from culture into civilization. And then came the great purifying trek that we call world war ... The

German landscape was almost completely preserved, the healing unity between the nature of our homeland and that of the human soul. (*DL*, 20)

Germany after 1918, Schreyer insists, has won back the potential of *Kultur* – that is the creative interplay between the German people and their landscape. Schreyer examines the artistic aspect of this in his volume *Sinnbilder deutscher Volkskunst* [*Symbols of German Folk Art*] (1936) in which he analyses folk art as belonging to the *Volksgemeinschaft* and as having the task to purify man by making him more aware of the culture he has lost. The images of folk art form the inwardly natural world of the *Volksgemeinschaft*. He develops his argument by identifying typical national or racial characteristics shown in art. That of the German people, as we have seen, he claims to be an indestructible longing for light. The final words of the book show how close he comes to acceptance of theories and practices not far removed from National Socialism: 'Flamelike, the power of the soul, made visible, rages full of light, upwards from the community, drifts across it, in front of it and leads it on. And the community, one people, follows the flag, follows an image and fulfils the meaning of the image.'[21] He was more specific in *Deutsche Landschaft* where we can read:

The soul can read the script of the stars. The voice of the invisible leader is calling. The community comes to order at his inaudible call. The people moves on following his law. The leader calls us to obey. The law of the people only fulfils itself when it follows his law. (*DL*, 21)

The German world lives, he claims, when it wanders through its landscape and rediscovers its sense of community, helped along by art, music and literature to appreciate the harmonic principles in nature, which he quickly claims as the traces of God.

Already in his dissertation on 'Literarisches und künstlerisches Urheberrecht' ['Literary and artistic copyright'] (Leipzig, 1910), Schreyer had tried to establish the conditions behind a prestabilized world system, not least in his experimental use of a prose style based on triadic structures. In the 1920s this concept led him to work out harmonical problems with the modern Pythagorean Hans Kayser, with whom he founded the Vereinigung mystischer Literatur [Association of Mystical Literature] and whose books he owned and used frequently as analytical models in his various statements on the balance in all creation and man's particular position within it. Schreyer's drawings such as the *Skizze für ein Marionettentheater* [*Sketch for a Puppet Theatre*] (1920) and *Struktur des himmlischen Menschen* [*Structure of Divine Man*] (1924–7) show man at the centre point of creation, surrounded by form-giving forces. In both, the relationship between microcosm and macrocosm is made obvious. Here Schreyer's holism suggests a path towards a more Christian view of the universe than the call of National Socialist Blood and Soil ideology. The link between the natural world and the religious-spiritual world came in particular through his study of the fifteenth-century German mystic Nicholas of Cusa with his vision of 'Coincidentia Oppositorum', and which he worked out together with Oskar Schlemmer in the Weimar Bauhaus together with the implications of the essay by Heinrich von Kleist *Über das Marionettentheater* [*On Puppet Theatre*].[22] In all these instances, Schreyer found confirmation of his belief that the natural world can reveal an ideal human world, one that for him would find expression as a *Heilige Weltschau* [*Divine World Revelation*], the title of a 306-page manuscript dated 1952. Here the traces of God revealed in nature are set alongside the 'images of God in creatures that are bound to him in spirit'.[23] Art, and particularly abstract Christian art, is now seen in terms of its light-bringing qualities, a theme that was central to both his art and his interpretations of art after he had overcome the temptations of Nazi ideology. Hence we find such statements as 'All of nature is one secret and obvious revelation of light.'[24] It is tempting to examine some of the many experimental statements Schreyer makes in his effort to convince the reader of the truth of his vision and faith. Many of them refer whole systems back to basic shapes or musical patterns or harmonies of proportion and colour, and these hark back to lectures he gave in the Sturmschule and the Weimar Bauhaus. His struggle in the early 1930s was based on his extensive knowledge of various mystical systems and of Hans Kayser's harmonical experiments to produce blueprints of physical and human universes, for these went further than the Blood and Soil ideology. Nature becomes one of many features within an all-embracing categorizing theory to which are fitted various modes of expression – art, music, architecture, human behaviour, etc. Schreyer's way to Christianity, and in particular Roman Catholicism, owed much to the contemplation of natural landscape, not as an aesthetic experience but as a series of signs to humanity, and in particular German humanity. Nature, he found, gains an ecological dimension when it culturally involves and 'purifies' man.

The most explicit statement of allegiance to the National Socialist cause comes in the essay 'Vom Landschaftserlebnis unserer Malerei' ['On the representation of landscape in our painting'] (1935).[25] Here Schreyer analyses four paintings by Werner Peiner, and we read in his introductory remarks:

> No art can be posited that does not construct its figures with the secret forces, which does not create life from the soil of our homeland. Landscape painting has in addition to this the special task of proclaiming in the work of art the energies of the soil of our homeland, its particular moulding, the secret that binds our soil and German man indissolubly together. For landscape is more than the outer face of nature which surrounds us with the sky, mountains, wood and meadows and all that is beautiful. Landscape is even more than the life that flows up towards us in continuous renewal from the soil. The forces of our human being are conjoined in a similar way in the landscape with the soil of our homeland. Landscape emerges in a communal unfolding of the forces of blood and soil. Landscape painting should proclaim these forces, their unfolding and their inner union, in which they alone can realize their own existence.[26]

That must seem clear evidence of Schreyer's involvement with party-inspired aesthetics, but worse is to come: 'Perhaps this realization is proclaimed in the most impressive way in the painting *German Soil*. This painting was granted the honour of becoming the property of the leader and Imperial Chancellor Adolf Hitler.'[27] A description of the picture and of three others follows, culminating in the claim:

> In such an immense tension between nature and man, between blood and soil, the life of the people persists, and this tension produces a resonance, a harmony of a wonderful kind that man can only hear within himself whenever he listens attentively to the life of his native landscape.[28]

This gesture of linking the inner face, as he calls it in some of his theories on art, to the landscape of one's immediate home environment becomes a key to understanding Schreyer's poetry in particular and his works in general. This may be summed up in the concept of translucency which I have analysed in the chapter on poetry in my monograph.[29] To paraphrase that, all his works are searches for a synthesis of sound and colour, at the centre of which stands man, the summit of creation, as a 'broken being of light'. Art, literature and nature can help him to a new synthesis (the Expressionist 'New Man') and redeem him on to a higher level. Indeed, it is in his poetry that Schreyer's development towards a form of Christian ecological understanding of nature most clearly appears.

In the unpublished cycle *Meerfahrt,* which bears the probably wrong date 1915, there are poems that must have been inspired by his frequent holidays on the island of Sylt. These were written at a time when, as a young man, he was still fascinated with the self-containment of natural phenomena, and where his enjoyment in playing with language, as so often in his works, is apparent. Experimentation on well-worn themes is developed in the Expressionist poetry he published in *Der Sturm* where he emphasizes the link between the forces of spring in nature and human passion. However, nature could be hostile, a warning, at times destructive, as can be seen in many stanzas of the huge unpublished cycle that he read publicly after 1945 entitled *Die menschliche Elegie. Gesänge zwischen Traum und Tag* [*The Human Elegy: Songs between Dream and Day*], but by then his poetic self was sometimes racked with despair, sometimes buoyed up on hope. Typical would be the prayer-like formula 'I am the stone. I am the way . . . Smash this stone and give only your form.' Furthermore, apocalyptic landscapes and the intervention of angels are never far away in his writings during and after the Second World War.

More typical of the 1930s to 1950s are poems, many of them sonnets, that interpret the voices of nature, and where nature is interpreted as a mediatory force. Another type of poetic statement pervades the forty-eight cantos of the *Menschliche Elegie* where more abstract statements of his intentions are often expressed.

To sum up, Schreyer first understood nature to reveal a spiritual reality behind everything. He developed a fascination with its independence and moved on to experimental Expressionist attempts to capture its meaning in linguistic form. Schreyer later used nature to support a National Socialist vision of Nordic man, but came to realize how its messages supported a quite different form of 'purification' from that of a supernatural homeland where man could discover for himself the light of religious experience. In nearly all of his pronouncements referring to nature, Schreyer implied a general ecological application for man, although ironically the flirtation with National Socialist ideas of racial purification may even have helped him on his way towards his vision of the purification of man in a mystic, Christian sense. If there is any doubt that Schreyer was led astray for long, this was corrected by the publication on 6 August 1939 of his essay 'Warum lieben wir die Natur?' ['Why do we love nature?'] in *Schönere Zukunft*,[30] a Viennese periodical, a reprint of the chapter 'Christliche Naturbetrachtung' ['Christian contemplation of nature'] from Kurt Ihlenfeld's *Die Stunde des Christentums. Eine deutsche Besinnung* [*The Hour of Christianity: A German Reflection*] (Berlin, 1938). For, just six years after *Deutsche Landschaft* and only three after 'Vom Landschaftserlebnis unserer Malerei', all hint of National Socialist ideology has gone. The litany of answers to the question as to why we love nature is a totally committed and obedient religious statement. The spiritual reality Schreyer eventually found confirmed in nature was one of adoration and identification; 'a victory over the demons that permeate and encircle our life on this earth'.[31]

Notes

[1] Lothar Schreyer, *Über die Grenzen der Natur*, 46. MS (293 pages), Deutsches Literaturarchiv, Marbach.

[2] Daniel Gasman, *The Scientific Origins of National Socialism: Social Darwinism in Ernst Haeckel and the German Monist League* (New York, London, 1971).

[3] Ibid., 165.

[4] Adolf Hitler, *Secret Conversations 1941–1944* (New York, Farrar, 1953), 43.

[5] Gasman, *The Scientific Origins of National Socialism*, 170.

[6] These include *Die Lehre des Jakob Böhme* (Hamburg, 1923), an edition of *Vom dreifachen Leben des Menschen* (Hamburg, 1924), a selection of texts in *Deutsche Mystik* (Berlin, 1924) and the articles 'Jakob Böhme und das Wort' in *Deutsches Volkstum*, 26 (1924), 473–8 and 'Jakob Böhme' in *Theosophie*, 13 (1924–5), 1, 18–32; 2, 118–33; 3, 196–211.

[7] Lothar Schreyer, *Deutsche Landschaft. Eine Naturbetrachtung* (Hamburg, 1932). Subsequent references to this work will be indicated by the abbreviation *DL*.

[8] For a full list consult the bibliography in Brian Keith-Smith, *Lothar Schreyer* (Stuttgart, 1990).

[9] Quoted from an unpublished manuscript in Deutsches Literaturarchiv, Marbach.

[10] Lothar Schreyer, 'Pferde', *Herdfeuer* (March 1929), 35-6.

[11] Ibid., 35.

[12] Ibid.

[13] Lothar Schreyer, 'Ausblick auf eine heroische Kunst', *Deutscher Wille* (August 1934), 214-18.

[14] Ibid., 215.

[15] Lothar Schreyer, *Die bildende Kunst der Deutschen* (Hamburg, 1931), 11.

[16] Ibid.

[17] Lothar Schreyer, *Meister der Graphik* (Hamburg, 1948-9), 217.

[18] In Lothar Huber (ed.), *Franz Werfel* (Oxford, 1989), 96-7.

[19] Lothar Schreyer, *Verantwortlich* (Hamburg, 1922).

[20] Ibid., 4-5.

[21] Lothar Schreyer, *Sinnbilder deutscher Volkskunst* (Hamburg, 1936), 190.

[22] Brian Keith-Smith, 'Heinrich von Kleist's *Über das Marionettentheater*', *New German Studies*, 12, 3 (1984), 175-99.

[23] Lothar Schreyer, *Heilige Weltschau*, unpublished MS (1952), 7.

[24] Ibid., 79.

[25] 'Vom Landschaftserlebnis unserer Malerei', *Herdfeuer*, 10, 2 (1935), 469-72.

[26] Ibid., 469.

[27] Ibid.

[28] Ibid., 471-2.

[29] Keith-Smith, *Lothar Schreyer*, 245-74.

[30] Lothar Schreyer, 'Warum lieben wir die Natur?', *Schönere Zukunft*, 16, 45 (1939), 1151-2.

[31] Ibid., 1152.

Part II
Contemporary Trends

5 • Ecological Thought and Critical Theory

Ingolfur Blühdorn

The works of the Frankfurt School of critical theory, above all Horkheimer and Adorno's *Dialectic of Enlightenment*, undeniably belong to the most influential philosophical and sociological writings of the twentieth century.[1] Yet if one tries to make a connection between contemporary ecological thought and this particular current of western post-Marxism, one is immediately confronted with the objection that, despite some obvious thematic overlaps, the Frankfurt School has had little direct influence on the theory and practice of the new social movements, including the ecology movement. Hülsberg complains about the 'coded language' of the Frankfurt School which made their thinking inaccessible to the protest movements at the grass roots of society.[2] Eckersley's criticism is that 'their overriding objective was the liberation of "inner" rather than "outer" nature', and that their protest against the domination of nature is ultimately anthropocentric rather than ecocentric.[3] Dobson comes to the conclusion that critical theory, instead of providing Greens with a theory of social change, 'gives us . . . a sophisticated theory suggesting its impossibility'.[4] Needless to say the Frankfurt School has never been concerned with ecological problems in the strict sense; equally, it is true that the evolution of political and social theory has moved beyond Horkheimer and Adorno. Arguably, they were the last theorists of modernism, and their successors entered the realm of postmodernism with its multiple discourses. However, it is precisely their wider understanding of 'nature', as well as their position on the threshold between modernism and postmodernism that ought to make the works of the Frankfurt School interesting for contemporary ecological theorists. These works can be of immense benefit in conceptualizing and theorizing the current state of the ecology movement and the related eco-theoretical debate.

Both the movement and its theorists are presently going through a phase of profound transformation. The thesis to be explored in this chapter contends that the most fundamental patterns of thinking which have dominated ecological thought for the last three and a half decades are gradually dissolving. This affects in particular the traditional concept of 'nature' which, in the 'age of its technical reproduction', loses its status of being 'given to us'.[5] Furthermore, this affects the social experience of the nature–civilization divide, raising hopes for an end to the suffering caused by mankind's civilizatory alienation from nature. Borrowing a term coined by Wolfgang Welsch with regard to the theory of post-modernism, one could say that green political thought is on the verge of completing its 'Trauerarbeit der Moderne' [the mourning process of modernity].[6] In philosophical thinking and social theory, which also have a long tradition of discussing the relationship between mankind and nature, the transition from modern to post-modern patterns of thinking has long been completed, notably through the works of the Frankfurt School of critical theory and the criticism they provoked. It is for this reason that critical theory is qualified to play the role of midwife as ecological thought enters a radically new phase of its evolution.

The argument of this chapter is developed in three stages. In the first part the emphasis is on describing the current state of the eco-movement and of ecological thought. A state of crisis is diagnosed, which is marked – and caused – by the radical plural-ization and democratization of the concept of 'nature'. Ecological thought is ill-prepared for this development because, up to the present, it has left the relationship between mankind and nature, particularly in its historical dimension, hopelessly undertheorized. Part two turns to critical theory for assistance by analysing its concepts of 'nature' and 'reconciliation'. In particular, Adorno's differentiation between 'first nature' and 'second nature' is regarded as a valuable stepping-stone on the way towards a new understanding of nature and our relationship towards it. The concluding section considers whether contemporary ecological thought might, with the help of critical theory, transcend critical theory itself and perform its own *gesellschaftstheoretische Wende* [transformation into social theory].

Ecology and crisis

The diagnosis of a crisis of the ecology movement and of ecological thought is contentious inasmuch as the former – particularly in Germany – has undoubtedly reached an unprecedented degree of organization, public support and operational efficiency,[7] while the latter has generated an immeasurable amount of academic literature adding eco-dimensions to every conceivable area of research. It seems paradoxical that the overwhelming success of the green movement and of ecological thought should coincide with their crisis. Yet as much as this success seems to prove wrong the assumption of a crisis, it may be taken as an indicator of its reality. Everybody and everything has turned green, evidencing a rapid pluralization and differentiation of the concept of environmentalism. A multitude of green discourses has evolved, but their common self-assessment as ecologically concerned in no way guarantees the possibility of their communicative interaction. What is described here as the crisis of the ecology movement and of ecological thought is the result of three basic weaknesses which have so far either not been perceived or been taken as curable diseases. The first of these weaknesses concerns the inability of ecologists to present a convincing theoretical model of the desired society realizing the dream of civilization's reconciliation with nature. The second is the lack of a strategy which could bridge the gap between the status quo and the desired future. And the third weakness is the remarkable ambiguity of what is at the bottom of all ecological thinking and campaigning: environmental concern. With regard to the social and political vision of ecologists it seemed at one time to be evident that the whole of capitalist and industrial consumer society had to be replaced by a different model of social and economic organization because human civilization had become alienated from nature and was destroying its own basis of regeneration and reproduction. Pacification, reconciliation, living in harmony with nature were the goals, democratic participation, the control of technology and the limitation of consumption key features of the ideal society. However, ecological thinkers tended to confuse what they perceived as the ecological bankruptcy of the dominant civilizatory model with its overall state. They overlooked the enduring strength of the constitutional model of a socially controlled capitalist consumer society.[8] Also, they were unrealistic about public interest in democratic participation, they underestimated the emancipatory and civilizatory potential of technological progress and failed to present effective incentives as well as convincing criteria for the reduction or social redistribution of wealth. Only the decline of socialism made ecologists admit that, rather than being 'neither right nor left', some kind of socialist society, however ecologically modified, had been their vision and ideal. Political developments in eastern Europe were the international referendum against any such vision. Bereft of its long-term perspective and unable to make up for the loss, the ecological identity was reduced to reformism aimed at the greening of consumerism. Even in this area successes turned out to be modest. Despite all criticism of material superabundance and all campaigning against further economic growth, the continued rise in consumer demand established itself as a key anthropological feature claiming a quasi-natural status. Talk of post-materialism had obviously raised premature hopes. The desire for active political participation, on the other hand, turned out to reach its limits where the conditions for personal material wealth had been fulfilled. Germany is a prime example of social movements adopting forms of representational protest[9] and the Green Party moving away from its grass-roots ideals, not just because they proved ineffective, but also because not enough 'roots' were available to support the Greens.[10]

With regard to the second weakness, it has often been pointed out that the eco-movement and ecological thought still need a convincing strategy for the transition to environmental sustainability.[11] Furthermore, ecologists have so far failed to identify a suitable subject of such social change.[12] At one time, the revolutionary eco-movement, which derived its identity and cohesion from eco-fundamentalism, seemed to be a promising agent of change. Within the framework of New Politics the movement and its organizations have indeed become key actors in the political arena;[13] there is, however, no evidence of an intrinsic link between the dispersal of political authority and the achievement of environmental compatibility. Far from giving rise to a new social consensus,[14] the green issue is itself subject to a process of diversification and fragmentation. 'In view of the irreversible differentiation of society', green radicalism has become anachronistic and has been branded as 'parasitic'.[15] Controversial views of the ecologically necessary and desirable replace the simplistic strategic perspectives of eco-fundamentalism and lead to disintegration of the eco-movement as a political actor.

Although ecological pragmatism has been hailed as the remedy for all green strategy problems, leading figures of so-called realism overlook the significance of fundamentalism's 'somewhat arbitrary metaphysical foundations'as a motivating force.[16] Ecological thinking has – apart from some reactionary authoritarian currents – mostly been embedded in the larger context of emancipatory thought.[17] Rational thinking and free will have been regarded as the tools for identifying the flaws in the current socio-economic system, for sketching out the ecological alternative and for managing the transition from one to the other. Exclusive reliance on rationality and voluntarism, however, made heavy demands on human beings. As driving forces they depended on the assistance of fundamentalism's metaphysical 'element of non-negotiability'.[18] The end of this non-negotiability invariably leads to hesitation or disorientation. What is to be protected or achieved? And why? Borrowing the words of Hubert Weinzierl, a long-standing spokesman for the ecology movement in Germany, one might say that the eco-movement is 'less than ever of one mind on the question of what . . . (they) want to preserve at all'.[19] This situation is further complicated by the experience that all action taken for the environment seems to effect the opposite of what it intends to achieve. However much further control over nature is resented, environmental activists and authorities, from local up to global level, are not liberating nature from the grip of human exploitative rationality, but are quantifying, categorizing, mapping, administering and bureaucratizing it in order to ensure that the most advanced techniques of protection may be applied. Given the global impact of human civilization which affects even the most remote spot of nature, an ever-stronger human input is needed to simulate in any chosen reserve the absence of anthropogenous influences. Increased interest in the natural environment accelerates rather than slows down the process of its appropriation. What is missing in green thought is not only a 'historical account of our relationship with the natural world',[20] but, more generally, a model of history which, without falling back into natural determinism, relativizes confidence in *Machbarkeitsdenken*, in the feasibility of an ecological future and of society and history in general.

The third weakness of ecological thought concerns its incentive. Beyond the concern for human health and fear of the destruction of the natural basis of life, the ultimate incentive for the eco-movement and its theorists is the experience of human alienation from nature and the desire for reconciliation with it. This ultimate concern places them firmly into the long tradition of the European movement of the Enlightenment which has been discussing the nature–humanity divide for over two hundred years. The subjective experience of alienation is hard to quantify empirically. It is, however, evident that this social experience must, looking back-wards, decline in proportion to the dissolution of the consensus on what is to be protected and why it is worth preserving, and, looking forwards, to the dissolution of the social ideal of the reconciled condition. The decline of the social movements and the de-radical-ization of green politics which, provocatively, has been called *The Fading of the Greens*,[21] reflect the decline of social suffering arising from the nature–society divide, and thus the loss of incentive for ecological thought.

The root of all three weaknesses is that from its inception in the 1960s, ecological thought has developed on the questionable assumption that civilization and nature are two clearly distin-guishable and separate entities which have built up a mutual opposition. It was assumed that the tension between the two poles was man-made, and hence mankind was regarded as morally responsible. For the benefit of both a freely developing nature and a human community liberated from the perversions of current civi-lization, this tension had to be resolved. In the recent literature the emphasis is clearly on the liberation of nature. The ecocen-tric current of ecological thought, regarding itself as 'the most comprehensive, promising, and distinctive approach ... in eco-political theory', confirms and reinforces the dichotomy between nature and civilization.[22] Putting all its energies into conceptual-izing nature's intrinsic value, this current has contributed significantly towards steering ecological thought to its current posi-tion of unmanoeuvrability. Focusing on the Other of society, trying to clarify humanity's place in nature and seeking to embed civi-lization into it, in other words taking nature as the static starting-point for all efforts at reconciliation, the ecocentric current is distracting attention from the crucial challenge to analyse the social, political and ideological function of nature *in society and for it*. In the age of postmodernity nature has become a socially negotiable concept rather than a pre-defined condition 'out there'. Ecocentrics refuse to realize that, at the peak of the ecological debate, the split between nature and civilization has lost its social and political reality. If it was the postmodern condition which gave rise to the crisis of the eco-movement and ecological thought, a postmodernist reassessment of the nature–society relationship has to be the starting-point for its solution.

Critical theory on nature and reconciliation

Environmental deterioration is certainly not one of the primary concerns of the works of critical theory. However, various analysts[23] have noted that long passages of Horkheimer's *Eclipse of Reason* and of Horkheimer and Adorno's *Dialectic of Enlightenment* read as if they had been written by ecological thinkers. The most obvious area of overlap between critical theory and ecological concern is their common critique of the Enlightenment, which is assumed to have alienated mankind from nature and whose tool, instrumental reason, has mutilated both outer nature and mankind's inner nature. For critical theory and ecological thought, the painful split between nature and humanity gives rise to the desire for the liberation of nature and reconciliation with it. Despite all scepticism with regard to the success of the Enlightenment, both critical theory and ecological thought firmly rely on reason itself as the prime diagnostic and therapeutic tool. Horkheimer and Adorno, however, were mainly concerned with the realm and the possibilities of philosophy. They restricted themselves to 'unshackling independent *thought*' which they considered as 'the sole way of assisting nature' [my emphasis].[24] In their view, reconciliation with nature would not become social reality, but would reside exclusively in art and philosophy. Ecological thought, on the other hand, and all the social movements in the context of which it evolved, are far more demanding of, and confident in, the self-correcting powers of human rationality. They assume that theoretical recognition can and ought to be immediately translated into social and political practice. The notion of taking control of developments, of planning and actively effecting social change and the liberation of nature is essential for social movements and ecological thought. This infinitely stronger confidence in human rationality is the point from where the significant difference between ecologists and early critical theorists unfolds. An analysis of critical theory's concept of nature and of the relationship between nature and its counterpart, human rationality, reveals how critical theory viewed the possibilities of sketching out a utopia of the reconciled condition, of developing a practical strategy bridging the gap between this vision and the status quo, and finding a suitable incentive or driving force for the transition.

Critical theory's most comprehensive discussions of the meaning and significance of nature can be found in the *Dialectic of Enlightenment* and in Adorno's later *Negative Dialectics*.[25] Neither of these books uses a homogeneous concept of nature. Depending on the context, the term adopts different meanings. It cannot be explored here in all its dimensions, yet with regard to today's ecological thought two aspects are particularly important: one is the idea of nature as having been created by the human subject rather than having existed prior to it, the other is the distinction between 'nature' and 'second nature'. Horkheimer and Adorno conceptualize the history of mankind as the history of increasing domination of nature. At the beginning of this history was the split of the *Sein* [Being] into the Self and the Other. The birth of the Self was followed by the need for Self-preservation *vis-à-vis* the Other, *vis-à-vis* nature which, by this very act, was constituted as the paragon of *Fremdbestimmung* [heterodetermination] obstructing human self-determination. From its very beginning, the relationship between the Self and the Other, i.e. the human subject and nature, has always been a relationship of one dominating the other: 'the essence of enlightenment is the alternative whose ineradicability is that of domination. Men have always had to choose between their subjection to nature or the subjection of nature to the Self' (*DE*, 32). It is important to note that for Adorno the idea of nature did not exist prior to the idea of the Self and has no reality independent of it. Hence, nature *is* or *appears as* 'always the same' and as nothing but 'a substratum of domination'; 'this identity constitutes the unity of nature', which was presupposed 'as little as the unity of the subject' (*DE*, 9). Although, however, the unity of nature is a product of the human mind rather than being presupposed to it, it is assumed that the establishment of nature and the human Self (object and subject) was preceded by some kind of mythical condition which existed *a priori,* and a knowledge about which – if it were possible – would reveal something about the potential qualities of the Other if it had not been galvanized into one unity under the concept of nature. For Adorno it is the central aim of all *philosophy of Being* to understand what the Other – which for reasons of simplicity and in line with the common use of language he often simply calls nature – could be like, independent of what human perception makes of it. The *philosophy of Being* 'seeks to breach the walls which thought has built around itself, to pierce the interjected layer of subjective positions that have become a second nature' (*ND*, 78f.).

What does he mean by 'second nature'? In the *Dialectic of Enlightenment* the distinction between nature and second nature is only hinted at, but it is fully elaborated in the *Negative Dialectics*. Horkheimer and Adorno see all historical progress as the refinement of human domination over nature. The ideal of human rationality and of the Enlightenment is 'the system from which all and everything follows' (*DE*, 7). What Horkheimer and Adorno call the 'system' is what human rationality erects with the aim of self-preservation and the domination of nature. The more successful instrumental reason is regarding the emancipation of the human Self, i.e. the more the non-rational forces of nature come under control, the more complex does this system become. This system, society, transforms into second nature:

> What is truly θέσει – produced by a functional context of individuals, if not by themselves – usurps the insignia of that which a bourgeois consciousness regards as nature and as natural . . . The more relentlessly socialisation commands all moments of human and interhuman immediacy, the smaller the capacity of men to recall that this web has evolved, and the more irresistible its natural appearance. The appearance is reinforced as the distance between human history and nature keeps growing . . . (*ND*, 357f.).

The system designed for the domination of nature becomes so complex that it becomes a source of *Fremdbestimmung* itself. It is the dialectic of the Enlightenment that 'there is a universal feeling, a universal fear, that our progress in controlling nature may increasingly help to weave the very calamity it is supposed to protect us from, that it may be weaving that second nature into which society has rankly grown' (*ND*, 67).

Nature and second nature are identical inasmuch as they both encroach upon the human struggle for self-preservation and self-determination. The identification of nature with second nature is unavoidable, yet the confusion of θέσει and φύσει is fatal because it entirely and irreversibly obstructs the view of nature as it might be irrespective of human perception. In order to remind us of this confusion, Adorno calls the false image the human mind creates of nature a *Verblendungszusammenhang* [delusive context] which tends to become an *universaler Verblendungszusammenhang* as the Enlightenment leaves nothing behind which is not penetrated by human rationality. As this delusive context is continuously being produced and reproduced by every cognitive act of the human mind, Adorno makes the following point:

> No less delusive [than the delusive context itself] is the question about nature as the absolute first, as the downright immediate compared with its mediations. What the question pursues is presented in the hierarchic form of analytical judgment, whose premises command whatever follows, and it thus repeats the delusion it would escape from. (*ND*, 359).

This is Adorno's criticism of the 'bourgeois ideal of naturalness' (*DE*, 31). The bourgeois concept of nature pretends to be speaking about the radically different, from which it hopes to derive ethical and political imperatives, while in reality it only captures a tamed and fully calculable nature. Where human reason is reflecting on nature, it is, without admitting it, only speaking about itself, about the 'spirit as the second nature' (*ND*, 356). Therefore Adorno rejects the bourgeois concepts of nature as being ideological. At the same time he is aware that the human mind can only capture second nature and that conceptualizing original nature is a contradiction in terms.

In Adorno's thinking, the fact that the *Verblendungs-zusammenhang* tends to become universal does not mean that nature disappears altogether. Its last and perhaps only manifestation is the force driving all efforts to dominate nature and thereby preserve the Self. This effort is the very principle of human reasoning and cannot be stopped unless reasoning itself stops. Adorno calls this the *Naturverfallenheit*[26] of mankind which 'cannot be separated from social progress' (*DE*, xiv). '*Naturverfallenheit* materialises as the subjugation of nature, without which spirit does not exist' (*DE*, 39).[27] Human reasoning cannot escape its *Naturverfallenheit*, yet as Adorno believes, it can become aware of it: 'Thinking, in whose mechanism of compulsion nature is reflected and persists, inescapably reflects its very own self as nature which has lost awareness of itself – as a mechanism of compulsion' (*DE*, 39).[28]

In the *Eclipse of Reason* Horkheimer demands programmatically that nature and reason must be reconciled without being equated with each other.[29] The *Dialectic of Enlightenment* and *Negative Dialectics* leave no doubt that such reconciliation will not be possible. Whatever we describe as nature and independent of human reasoning is in reality second nature. What may be considered as nature's last retreat is not accessible to reason. Factually, nature and the nature–humanity dichotomy have been abolished. For reasons of clarity and in order to avoid all ideological identification of nature with second nature, Adorno demands 'to see all nature, and whatever would install itself as such, as history' (*ND*, 359). This means adopting a completely social concept of nature. Despite the practical abolition of nature, however, Adorno insists that 'human history, the history of the progressing mastery of nature, continues the unconscious history of nature, of devouring and being devoured' (*ND*, 355). Nature does not disappear totally. It remains the law which drives and binds all historical and social progress. Turning to Marx, Adorno argues: 'Even if a society has found its natural law of motion . . . natural evolutionary phases can be neither skipped nor decreed out of existence ...' (*ND*, 354). The reproach of contemporary ecocentrics, that critical theory 'denies the fact of humanity's embeddedness in nature',[30] Adorno would have firmly rejected. Since the idea of nature neither can nor will ever disappear altogether, the hope that humanity or civilization might be reconciled with nature must always persist. So far as philosophy is concerned, at least, nature and reconciliation will maintain their importance as, in the Kantian sense, *regulative ideas of reason.*

Beyond critical theory

How can these findings contribute to the conceptualization and solution of the particular problems which cause ecological thought's current crisis? This crisis was earlier described as consisting of three elements: the lack of a convincing vision of a reconciled society existing in harmony with nature, the absence of a practical strategy to take society from here to there, and the crisis of motivation, as the experience of alienation and the desire for reconciliation seem to fade away. A close look at critical theory reveals that

Adorno and Horkheimer, despite never having been directly interested in environmental matters, anticipated these problems: it is so difficult to sketch out a vision of a reconciled society because there are 'no concepts available which could capture the state of reconciliation, whose idea appears [if at all], *ex negativo*, on the horizon of art and philosophy'.[31] It is so difficult to outline a practical strategy of transition firstly because human rationality is incapable of conceptualizing the radically different, and secondly because of humanity's *Naturverfallenheit*. Ecological thought, despite firmly believing in the limitedness of human capabilities, grossly overestimated the capacities of human rationality. It shares the *Machbarkeitsdenken* of the civilizatory model it attacks. In that respect, it is pre-critical. Finally, it is so difficult to keep up the momentum of the ecology movement because the *universale Verblendungszusammenhang* blocks all access to nature. By establishing a second nature which denies the existence of nature, the progress of human reason makes the experience and conceptualization of the nature–society split increasingly impossible. Yet this experience of alienation and the resulting desire for reconciliation were the lifeline of the eco-movement. After the abolition of nature, however, ecological thought does not, as it might fear, become superfluous. On the contrary, as critical theory points out, the recognition of different concepts of nature as reflections of different stages of society's historical development puts all the more emphasis on the necessity of exploring their ideological content. This guarantees the continued need for an eco-movement and for an ecological debate.

In the ecological debate the idea of the *end of nature* is not entirely new.[32] In the more recent literature Gernot Böhme pointed out that any 'reference to nature as it exists through itself . . . is an illusion' and 'proves ideological inasmuch as it refers to an idea of . . . something invariable just at the point of time when – historically, probably irreversibly – it is dissolving'.[33] Böhme is therefore 'sympathetic to the demand to abolish the concept of nature altogether'.[34] In a similar vein Ulrich Beck noted the dissolution of the opposition between nature and society, and the evolution of a 'nature-society', brought about by the simultaneous processes of the 'socialization of nature' and the 'naturalization of society'.[35] Any such ideas of the abolition of nature, however, meet with immediate and manifold objections. In European cultures the most fundamental threat is probably that without nature there would be 'nothing but us' and we could 'no longer imagine that we are part of something larger than ourselves'.[36] Mankind would be left in a philosophical or cultural vacuum. The ubiquitous presence of nature in virtually all public discourse not only proves that there is an urgent need for such a larger context, but the growing interest in nature indeed seems to indicate the exact opposite of nature's abolition. From the scientific or empirical point of view there is the objection that human civilization is as far as ever away from the old ideal of controlling 'even the more remote side-effects of our . . . production facilities' and overcoming the 'contradiction between spirit and matter, man and nature'.[37] Natural catastrophes, human illnesses and the mere existence of death prove the persisting presence of nature. Economists, finally, make a clear distinction between 'natural capital' and capital derived from nature by humans. Without this distinction the whole debate about sustainability would not make sense, and neither would the demand for the 'internalization' of the natural costs of production.

However legitimate these objections may be, ultimately they miss the point of the thesis of the end of nature and cannot affect its general validity. Böhme points out himself that:

> Nature is always larger than mankind. Every improvement and every destruction of nature we cause, remains within its framework. . . . Furthermore, mankind's manipulation of nature is never unrestricted or unconditional. In some way man always depends on nature as a 'material', he always has to employ the existing laws of nature.[38]

Adorno, too, insisted on the continued presence and power of nature which manifests itself in the form of mankind's *Naturverfallenheit*. The important point in this context, and the reason for referring back to critical theory, is not so much the factual and total abolition of physical nature, but rather the abolition of the man-made concept of nature *as the Other* of mankind, as independent of it. Critical theory helps us to understand that physical nature has not just been completely transformed into human environment, but that our perception and idea of nature have never captured nature as independent of and distinct from mankind. On the contrary, nature has always been a socially mediated idea. The concept of second nature is very helpful at both the physical and the philosophical level. With horror, Robyn Eckersley notes that for critical theory, 'the unity of humans with nature is achieved by making it our artifact, by totally domesticating it.'[39] As an ecocentric Eckersley strictly rejects the idea of a second nature, but ultimately this is not a question of accepting or rejecting: it is a question of conceptualizing a historical reality.

However, Adorno and Horkheimer only pointed the way, for in the end ethical reservations prevented them from being consistent in their reasoning. The development they described they saw as a predominantly negative development. On the one hand this was the process of human emancipation and liberation, which could never be stopped or even reversed:

> we are the heirs, for better or worse, of the Enlightenment and techno-logical progress. To oppose these by regressing to more primitive stages does not alleviate the permanent crisis they have brought about. On the contrary, such expedients lead from historically reasonable to utterly barbaric forms of social domination.[40]

On the other hand this process of liberation was understood as the path into a new suppression of human needs which was seen as even more barbaric than in earlier stages of history, because this time suppression was inflicted on humans by humans. The fear that the negative side-effects of the process of enlightenment may outweigh its positive content, and that this development may be beyond human control, gives rise to the pessimistic outlook of critical theory. This fear has remained a key feature in German critical literature; witness Ulrich Beck's *Risikogesellschaft* [*Risk Society*]. In Horkheimer and Adorno's writings, however, the awareness that every increase in human self-determination is paid for with a signif-icant loss of something else is much stronger than in the works of contemporary theorists. When Horkheimer speaks of 'man's subju-gation',[41] when in the *Dialectic of Enlightenment* the authors state that 'Enlightenment is totalitarian' (*DE*, 6) and that it 'behaves toward things as a dictator toward men' (*DE*, 9), when the balance of all human achievements is struck with the statement that 'the fully enlightened earth radiates disaster triumphant' (*DE*, 3), whilst the demanded way of thinking resembles 'metaphysics at the time of its fall' (*ND*, 408), all this is the expression of critical theory's *Trauerarbeit* (mourning process) which obstructed the view of the positive potential of the discovery that nature is much less independent of mankind than the concept of nature as the Other would suggest. Adorno stretched the thinking of modernism to its ultimate limits, arguably even beyond them. He pushed the door open to postmodernism, but then left the next generation to cross the threshold. The essential inconsistency in Adorno's thinking is that he sticks to a model of linear modernization. His 'principle of dissolvent rationality' (*DE*, 6) is always directed towards the Other of the human mind, which is gradually penetrated until everything has been integrated into the system. Although Adorno points out that the replacement of nature by second nature is so complete that the *Verblendungszusammenhang* is practically universal, he holds fast to the distinction between nature and second nature, between the Self and the Other. Adorno needs this idea of modernization as a linear process because the unity of the human Self and of human reason – the very basis of modernist thinking – depend on the uniformity of the goal of *Naturbeherrschung* (domination of nature). If, however, the *Verblendungszusammenhang* really becomes universal, the metaphysical anchor of the unity of the subject, of uniform reason and the unity of nature, dissolves, and 'the principle of dissolvent rationality' turns against the 'spirit as the second nature' (*ND*, 356). Modernization then becomes reflexive.[42]

Postmodernism announced the pluralization of the subject and of reason, yet it forgot to announce the pluralization of nature, or – the other way round – the eco-movement and its theorists missed the advent of postmodernism and stuck to a concept of uniform nature which no theory can any longer support. In this, ecological thought is similar to critical theory, yet the latter was at least aware of its own inconsistencies and transformed them as a dialectic. Böhme indicates the step beyond Adorno when he demands that politics 'must stop treating nature as something extra-political' and instead start to negotiate 'which nature we want at all'.[43] In a similar way Beck points out that nature loses its status of being given to us and becomes 'a societal project'.[44] Once the idea of nature as the Other of humanity has been given up, once the concept of nature has been pluralized and recognized as a social concept, the 'transformation into social theory' (of ecological thought) has been completed, as, indeed, has the 'mourning process of modernity'. Then there is new scope for reconciliation which will be the reconciliation between partial forms of rationality with particular, socially negotiated ideas of nature. The pains of this transition are the cause of the current crisis of ecological thought.

Notes

1. See Theodor W. Adorno and Max Horkheimer, *Dialectic of Enlightenment*, tr. John Cumming (London, New York, 1979). Subsequent references to this work will be indicated by the abbreviation *DE*.

2. Werner Hülsberg, *The German Greens: A Social and Political Profile* (London, 1988), 9.

3. Robyn Eckersley, *Environmentalism and Political Theory: Towards an Ecocentric Approach* (London, 1992), 11.

4. Andrew Dobson and Paul Lucardie (eds.), *The Politics of Nature: Explorations in Green Political Theory* (London and New York, 1993), 207.

5. See Gernot Böhme, *Natürlich Natur. Über Natur im Zeitalter ihrer technischen Reproduzierbarkeit* [*Naturally Nature. On Nature in the Age of its Technical Reproducibility*] (Frankfurt am Main, 1992).

6. See Wolfgang Welsch, *Unsere Postmoderne Moderne* (Weinheim, 1987).

7. See Ingolfur Blühdorn, 'Campaigning for nature: environmental pressure groups in Germany and generational change in the ecology movement', in Ingolfur Blühdorn, Frank Krause and Thomas Scharf (eds.), *The Green Agenda: Environmental Politics and Policies in Germany* (Keele, 1995); Thomas Princen and Matthias Finger, *Environmental NGOs in World Politics* (London, 1994); Russell J. Dalton, *The Green Rainbow: Environmental Groups in Western Europe* (New Haven, London, 1994).

8. See Ulrich Beck, *Die Erfindung des Politischen* (Frankfurt am Main, 1993).

9. See Blühdorn, 'Campaigning for nature'.

10. See Joachim Raschke, *Die Grünen. Wie sie wurden, was sie sind* [*The Greens. How They Became What They Are*] (Cologne, 1993).

11. See Andrew Dobson, *Green Political Thought* (London, New York 1995); Joschka Fischer, 'Die Krise der Umweltpolitik', in Ulrich Beck (ed.), *Politik in der Risikogesellschaft. Auf dem Weg in eine andere Moderne* [*Politics in the Risk Society. On the Road to Another Modernity*] (Frankfurt am Main, 1991); John Ferris, 'Political realism and green strategy', introduction to Helmut Wiesenthal, *Realism in Green Politics: Social Movements and Ecological Reform in Germany* (ed. John Ferris) (Manchester, 1993).

12. See Robert E. Goodin, *Green Political Theory* (Cambridge, 1992) and

Wiesenthal, *Realism in Green Politics*.

13. See Princen and Finger, *Environmental NGOs in World Politics* and Dalton, *The Green Rainbow*.

14. See Beck, *Die Erfindung des Politischen*, and Ingolfur Blühdorn, 'Ecological consensus and anti-modernist simplicity', *Debatte: Review of Contemporary German Affairs*, 2 (1994), 64–85.

15. Wiesenthal, *Realism in Green Politics*, 38 and 216.

16. Ferris, 'Political realism and green strategy', 4.

17. See Eckersley, *Environmentalism and Political Theory*.

18. Ferris, 'Political realism and green strategy', 4.

19. Hubert Weinzierl, *Das grüne Gewissen. Selbstverständnis und Strategien des Naturschutzes* [*The Green Conscience*] (Stuttgart and Vienna, 1993), 11.

20. Andrew Dobson, 'Critical theory and green politics', in Dobson and Lucardie, *The Politics of Nature*, 90–209, (199).

21. Anna Bramwell, *The Fading of the Greens: The Decline of Environmental Politics in the West* (New Haven and London, 1994).

22. Eckersley, *Environmentalism and Political Theory*, 27.

23. See for example Thomas Link, *Zum Begriff der Natur in der Gesellschaftstheorie Theodor W. Adornos* [*On the Concept of Nature in the Social Theory of Theodor W. Adorno*] (Cologne and Vienna, 1986); Peter Cornelius Mayer-Tasch (ed.), *Natur denken. Eine Genealogie der ökologischen Idee*, two vols. (Frankfurt am Main, 1991); Eckersley, *Environmentalism and Political Theory*; Dobson, 'Critical theory and green politics'.

24. Max Horkheimer, *Eclipse of Reason* (New York, 1947), 127.

25. Theodor W. Adorno, *Negative Dialectics* (London, 1973). Subsequent references to this work will be indicated by the abbreviation *ND*.

26. In his English version of the *Dialektik der Aufklärung* [*Dialectic of Enlightenment*] John Cumming offers a range of translations for the German term *Naturverfallenheit*. None of them actually captures its meaning. The term does not say anything about nature or the state it is in, but rather something about mankind and human reasoning. It seeks to express that man is and remains a slave to nature and is, however much nature is subjugated, ultimately unable to free himself entirely from nature's power. In order to avoid misinterpretations I will stick to the German term.

27. Cumming translates: 'The decline, the forfeiture, of nature consists in the subjugation of nature without which spirit does not exist' (*DE*, 39). This translation is grossly distorting.

28. Cumming translates: 'Thinking, in whose mechanism of compulsion nature is reflected and persists, inescapably reflects its very own self as its own forgotten nature – as a mechanism of compulsion' (*DE*, 39).

29. Horkheimer, *Eclipse of Reason*, 122.

30. Eckersley, *Environmentalism and Political Theory*, 112.

31. Albrecht Wellmer, *Zur Dialektik von Moderne und Postmoderne. Vernunftkritik nach Adorno* (Frankfurt am Main, 1985), 19.

32. Edward Goldsmith, 'De-industrializing society' (first published in 1977), in *The Great U-Turn: De-industrializing Society* (Bideford, 1988), and Bill McKibben, *The End of Nature* (New York, 1989).

33. Böhme, *Natürlich Natur,* 22 and 115.

34. Ibid., 22.

35. Ulrich Beck, *Die Erfindung des Politischen* (Frankfurt am Main, 1993), 179.

36. McKibben, *The End of Nature*, 83.

37. Friedrich Engels, *Dialektik der Natur* [*Dialectic of Nature*] (1871), in Marx/Engels, *Werke*, 20 (Berlin (GDR), 1956ff.), 307–572, here 453.

38. Böhme, *Natürlich Natur,* 110.

39. Eckersley, *Environmentalism and Political Theory*, 90–1.

40. Horkheimer, *Eclipse of Reason*, 127.

41. Ibid., 105.

42. See Beck, *Risikogesellschaft*.

43. Böhme, *Natürlich Natur,* 24.

44. Beck, *Die Erfindung des Politischen*, 178.

6 • New Age Mysticism, Postmodernism and Human Liberation

Peter Thompson

In his *Critique of Hegel's Philosophy of Right*, Marx outlines his categorization of a tripartite unity of English economics, French politics and German philosophy and presents the first real systematic exposition of the concept of the *verspätete Nation* [belated nation] in Germany, putting its causation down to the objective weakness and consequent subjective vacillation of the middle class.[1] With the ecological movement of the 1970s and 1980s, however, we have a partial inversion of Marx's reflections on the Europe of 1843. English economics, in the form now of Friedmanite neoclassical liberalism, becomes increasingly dominant throughout Europe – including eastern Europe – yet in the field of the 'history of the future' it is the once radical and socially progressive French new middle class which retreats into the abstract philosophizing of post-structuralism and the German new middle class which develops a *political* dimension which is far ahead of the rest of Europe on the question of humanity and ecology. For once it would appear that Germany has not played the role of the *verspätete Nation* [belated nation], constantly following behind the crowing of the Gallic cock (in this case the deconstructionists).

However, as Blühdorn points out in the previous chapter, while the German Green Party has a very strong and increasingly pragmatic political profile, the green movement as a whole rests on a relatively undeveloped theoretical and philosophical foundation.[2] What I wish to do here is to look behind this apparent inversion for some of the structural reasons for the development of ecological consciousness at this particular point in German history and ask whether being far ahead in terms of green politics is the same thing as being essentially progressive in terms of the Enlightenment project of human liberation.

The Green Party famously described itself as 'neither right nor left but ahead'. However, it could be said that this slogan was a distorted reflection of the fact that ecology as an ideology is neither right nor left but confused. It is clear, for example, that certain forms of ecological, and more specifically ecocentric, thought can have some distinctly right-wing and irrationalist tendencies.[3] The reason for this confusion lies in a situation where the social element of the relationship between humanity and ecology is neglected and a clearly defined goal of social transformation or an agency for that transformation is almost entirely absent. Ecocentric thinking tends to deal with one of two levels; either the microcosmic or the macrocosmic. The one believes that all that matters is the atomic level, the behaviour of the individual, the other that there is a cosmic totality in which we are all but individual atoms.

It assumes that there is a given teleologically predetermined ideal whole, a primary nature, which must not be disturbed and is predicated on the unquestioned assumption that there is indeed a definable ecological crisis. This chapter will question both assumptions and posit that what is missing from the equation is an analysis of the dynamic dialectical relationship between micro- and macrocosm.

The position taken in this chapter assumes that there is indeed such a thing as totality and truth, but that these are dynamic and changing concepts, and that truth is in effect the search for, understanding and transformation of the totality, which is in turn the ultimate transcendence of partiality and *difference*. From this it should be clear that the approach is hostile to both postmodernist or so-called de-centred interpretative models which seek to posit that there is no such thing as the totality and that truth is essentially unknowable, as well as those which claim that the totality is all that counts and that the individual, or indeed the human species as a whole, is unimportant.

The question, however, is why this debate, which has tended to dominate philosophy and literary criticism recently, is relevant to the relationship between ecology and human liberation?

Important strands of ecological thought share distinctive characteristics with post-structuralist thinking, and some of the positions adopted by post-structuralists are replicated by many of the central ideologists of radical ecology (just as they are in radical feminism). Indeed radical ecology and radical feminism are seen as essentially one and the same thing by many deep green thinkers.

Were this sort of thinking limited to a few isolated individuals in both the postmodernist and radical green milieu there would be no real problem. However, just as with the ontological relativism of post-structuralism, the species relativism and anti-humanist episte-mology inherent in radical ecological thought is no longer a fringe intellectual activity but is beginning to be the new orthodoxy against which rationalists and materialists of all political hues, including many within the green movement, are forced to defend themselves.

Just as many postmodernist thinkers are forced by their neo-solipsistic conclusions to re-embrace the concept of spirituality and metaphysical determination, so are many green thinkers increas-ingly prepared to accept neo-metaphysical explanations for material phenomena. I shall show how in many ways, radical ecologists, despite their emphasis on 'otherness' and de-centred truth, accept the concept of a transcendental signified when it comes to ecology, and that as a result of their post-materialist *Weltanschauung* [ideo-logy] they are forced to view it as non-material reality. Of course, the most popular and common expression of this trend is the move towards New Age spirituality which is to be found in west European and North American societies, and which is increasingly finding expression in both literature and film. What is necessary is a materialist explanation of the crisis of materialism and the new triumph of spirituality, as well as a demonstration that postmod-ernist New Age thought and green spirituality are barriers to rather than the midwives of real collective human liberation. In demon-strating this contention, I shall take Rudolf Bahro's work as indicative of these particular ideas, with his emphasis on the redis-covery of the spirituality of the individual as the solution to the crisis of the totality.[4]

In a *Spiegel* article in 1992 Rudolf Bahro is quoted as saying that the ecological crisis has reached such a level that there is no longer any place for democratic decision-making about humanity's relation-ship with nature. What he had to say was based on three contentions:

1. We need a new theocracy.
2. There can be no democracy amongst a pack of wolves.
3. I am the prince of the ecological revolution.[5]

It is clear that one should not extrapolate from one example, and that Bahro is a figure who has always sought to provoke. However, it cannot be ignored that New Age spirituality and shamanism mixed with ecological authoritarianism and biological egalitarianism are, and have been since the late 1960s, common currency.[6] The reason I choose Bahro, therefore, is because he represents in extreme form a current of thought which is central to much green analysis. Also it cannot be argued, as Bahro has himself attempted to do, that his interviews have been misunderstood. If we look at his writings of the 1980s and 1990s, and in particular his book *Logik der Rettung* [*The Logic of Deliverance*],[7] we can clearly discern elements of eco-dictatorial thinking based on spiritualistic shamanism. Indeed the third section of that work is entitled 'ORDINE NUOVO auf spirituellem Fundament' in the full recognition that there already exists an Italian neo-fascist organization with the same name. His attempted pre-emptive defence against the charge of eco-fascism is that a new order is necessary and that those who refuse to accept such a necessity are to 'remain silent'. To move from the recognition of a necessity to positing the solutions he goes on to promote is a long step, however. Those solutions themselves only serve to reinforce the suspicions many had about his political trajectory after leaving the GDR. Statements such as 'If we really want to survive, we have to erect an efficient eco-dictatorship as quickly as possible' and 'the movement is the breeding ground for the princes of transformation' do little to assuage fears of the sort of society Bahro is promoting.[8]

Richard Rorty writes in another context that

> ... It is only when a Romantic intellectual begins to want his private self to serve as a model for other human beings that his politics tend to become anti-liberal. When he begins to think other human beings have a moral duty to achieve the same inner autonomy as he himself has achieved, then he begins to think about political and social changes which help them to do so. Then he may think that he has a moral duty to bring about these changes, whether his fellow citizens want them or not.[9]

What Rorty fails to recognize here is that the tendency towards a mistaking of their own personal desires for the needs of all humanity is an eminently bourgeois concept and not necessarily that far removed from the 'North Atlantic post-modern bourgeois-liberal neo-pragmatis(m)' (in other words social quietism) which he posits as being the best way forward for humanity. Ecology as an ideology provides the disillusioned radicals of the postmodern movement with the opportunity of re-creating a Derridean 'Transcendental Signified' whilst at the same time maintaining their positional purity.

This transcendental signified, however, resurrects the Kantian Sublime in the form of a neo-metaphysical god of nature or Gaia, but takes it further in a Schopenhauerian direction by emphasizing a spiritual rather than a social identity between the ecological self, i.e. the microcosm, with the ecological totality or macrocosm. As is well known, Schopenhauer posited that Kant's *Ding an sich* was identical with the will, that all being was the product of the individual will and that the individual will was identical with the cosmos. The idea that the only 'true reality' is the product of individuated perception rather than sensuous objective *a priori* existence puts the individual not only at the centre of society but of the universe and, more importantly, the universe at the centre of the individual. As Schopenhauer himself says:

> We have not regarded the manifold phenomena as the different species of the same genus which they are but have taken them for heterogeneous: that is also why there could not be a word to denote the concept of this genus. Hence I give the genus the name of its most admirable species, a nearer, immediate knowledge of which leads us to indirect knowledge of all other species.[10]

This name was, of course, *Will* and, as Schopenhauer goes on to contend, 'natural force is groundless, i.e. it lies right outside the causal nexus and the domain of the thesis of the ground and is perceived philosophically as the immediate objectivity of the will, which is entire Nature's In-itself.'[11]

In other words, the concept of Gaia and chaos theory is anticipated by Schopenhauer's identity of Will and natural force. Nature, in this scenario, becomes a transcendental *Ding an sich* which lies outside the causal nexus, denies the causal nexus and is ultimately threatened by humanity's supposedly mechanistic attempts to construct causality. The only way to apprehend the 'truth' of this reality is, to come to Nietzsche of course, through intuition rather than knowledge, through feeling rather than through thinking, through myth rather than science, through the beautiful rather than the true, through Being rather than through Becoming. Here Bahro echoes Nietzsche in his demand that scientists should give up their 'roles as secularized priests and become fully integrated human beings',[12] implying that a commitment to rational discourse and Enlightenment values is at the root of the dichotomy between man and nature.

The question is, however, why the rediscovery of the irrational occurs in the late 1970s and is becoming increasingly important today. The answer, I believe, lies in what Enzensberger in 1973 called the 'class character of the ecology debate'.[13] As he pointed out then, ecological destruction has always been present in the lives of ordinary working people. The decline in heavy industry reduced one source of pollution, the growth of private transport has led to other environmental problems. This means that pollution has become for many middle-class people emblematic of the worst problems they face in their increasingly insecure everyday lives even though, in overall terms, the level of urban pollution has fallen over the last twenty years. As Betz has pointed out, 'if the Greens were merely concerned with environmental issues they would have hardly received the intense ... domestic and international attention they have.'[14] And as Enzensberger once cogently remarked, it is only because pollution has impinged on the lives of the middle class and because they are unable to use traditional means of escaping it through purchasing power that it has become a prominent social issue. What has happened as a consequence is that, as with all middle-class issues, and in typical Rortian manner, the middle classes believe it is not a class or political issue but a universalist/non-political one which has nothing to do with the way society is organized but only with the way individuals behave.

In *Der Steppenwolf* Hesse has the Steppenwolf argue that the bourgeoisie is caught between hedonism und asceticism and that the class is too timid to choose. It cannot give up its individuality to asceticism and God, but neither can it accept the discomfort and intensity of hedonism. 'One can only live intensively at the cost of the self.'[15] What the new green middle class has done is to transform this dilemma into a virtue, and green ideology enables one to do so. One can live the ascetic, recycling and anti-consumerist self-denying life subordinating oneself to the needs of nature or Gaia rather than God (and feeling rather self-satisfied as a result), but one can also liberate oneself from traditional bourgeois, or rather petty-bourgeois morality in terms of personal relationships, sexuality, consumption of stimulants, etc., all of which are to do with the liberation of the individual rather than a class or humanity as a whole. In essence they live a life of hedonistic asceticism[16] which helps to maintain the balance between individual and cosmos which Hesse outlined, and avoids the need for social action. Indeed Bahro quotes Hesse's poem 'Stufen' ['Steps'] as an expression precisely of this unmediated dichotomy between individual and collective in which, out of extreme atomization, a new 'great tribe' will emerge.[17] As with Hesse, however, Bahro presents no evidence as to how this turn from the individual to the collective will come about other than through a recognition of individual spirituality and a return to the 'God-head'.[18] As Enzensberger points out, 'the bourgeoisie can conceive of its own imminent collapse only as the end of the world.'[19] As a class it conceives of its interests as identical with those of humanity.

The consequences of this are that a global or universal approach has to be adopted within which the primacy of individualism can be maintained. Because humanity is, necessarily, brought into the equation, however, ecology ceases to be a scientific appraisal and becomes a political programme supposedly no longer based on class but on universal needs. In this sense political ecology is the perfect postmodern polity because it welds the interests of the individual with the universal and yet appears to do so non-politically. It replaces the Marxist concrete universalism based on an analysis of class interest and conflict with a postmodern abstract cosmology based on an emphasis of the apparently positive nature of social atomization (Rorty's 'pragmatism'). Ecology emphasizes the essential unity of the particular, of *difference,* in the 'oneness' of cosmic totality.

Bahro also states that the world economy must be redivided and self-sufficiency must become the guiding principle. 'Small is beautiful' is based on the concept that small-scale changes are conserving and incremental. The reimposition of a decentralized and anti-industrial culture could, though, imply a return to organicist and irrational concepts of strict hierarchy and a 'recognition of necessity' as imposed by a 'prince' at the head of a feudal particularist agglomeration of eco-*Volksgemeinschaften.*

Enzensberger predicted precisely this sort of development some twenty years before Bahro:

> We must reckon with the likelihood that bourgeois policy will systematically exploit the resulting mystification (of commodity production) – increasingly so, as the ecological crisis takes on more threatening forms. To achieve this it only needs to demagogically take up the proposals of the ecologists and give them political circulation. The appeal to the

common good, which demands sacrifice and obedience, will be taken up by these movements together with a reactionary populism, determined to defend capitalism with anticapitalist phrases.[20]

As we can see, the calls for sacrifice and obedience are clearly discernible in Bahro's position. In terms of sacrifice the emphasis is put on asceticism, and in terms of obedience the social Darwinistic, even eco-fascistic elements of the metaphor of the wolf pack for humanity are clear.

But there is another element of obedience which needs to be dealt with, and that is the question of the role and position of humanity as a whole. Increasingly there has also been a trend towards biological egalitarianism or species relativism. That is to say, the contention that rather than being a special part of nature, humanity is no more and no less entitled to 'have dominion over the earth' than any other species.

Indeed this view can be found in many radical green ideologists as well as in German culture. Murray Bookchin and David Pepper have pointed out how writers such as Wendell Berry and James Lovelock have gone over to an essentially anti-humanist and anti-human stance in the name of ecology. Berry speaks of humans as 'the most pernicious mode of earthly being ... an affliction of the world, its demonic presence. We are a violation of Earth's most sacred aspects.'[21] Lovelock goes as far as to maintain that 'it is the health of the planet which matters, not of some individual species. This is where Gaia and the environmental movements which are concerned first with the health of people part company';[22] and that 'humanity will give way to "those species that can achieve a new and more comfortable environment ... a change in regime to one that will be better for life but not necessarily better for us."'[23]

In Germany too, this mood of absolute anti-human pessimism gained ground in the late 1970s. Botho Strauß, for example, in his *Groß und klein* of 1978 has the following monologue:

> Siehe, der Mensch wird
> abgehen von dieser Erde
> und aus sein in allen seinen Werken.
> Hinter ihm wird der Boden erröten in
> Scham und Fruchtbarkeit.
> Die Gärten und die Felder werden
> in die leeren Städte einziehen,

die Antilopen in den Zimmern äsen, und
in offenen Büchern wird der Wind zärtlich blättern.
Die Erde wird unbemannt sein und aufblühn.
Die gefesselte Hoffnung, befreit von jeglichen Propheten,
wird erlöst sein und in der Stille reichlich wirken.[24]

[Behold, man will depart this earth/ and be finished in all his works./ Behind him the soil will blush with/ shame and fertility./ The gardens and the fields will/ move into the empty towns,/ the antelopes will graze in the rooms, and/ the wind will gently leaf through open books./ The earth will be unmanned and will bloom./ Hope, everywhere in chains, will be liberated from all prophets,/ be saved, and in stillness do its plentiful work.

What is this 'gefesselte Hoffnung' this 'hope, everywhere in chains', which is to be 'liberated' from a human-based existence, and which exists apparently in abstraction from real human hope, other than some disembodied transcendental signified? This short verse embodies the contradiction between the liberation of humanity and the liberation of the earth from humanity. The essential difference must be between those who see the liberation of humanity as the precondition for the liberation of the earth from human exploitation, and those who see the liberation of the earth from humanity as the precondition for salvation.

This quotation, based on the Book of Revelation, was typical of the mood of the period. Allusions to the sinking of the *Titanic* and Udo Lindenberg and Günter Amendt's songs *Keine Panik auf der Titanic* [*No Panic on the Titanic*] and *Grande Finale* expressed the widespread feeling at the time that the apocalypse was upon us. The famous Shell study of 1981 showed that over 22 per cent of young people believed there would be increasing destruction of the environment, 21 per cent believed in the *Weltuntergang* [end of the world] and 20 per cent that there would be another world war (this time nuclear).[25] But perhaps most importantly, all of this was seen in quasi-religious terms. Both the ecological problem itself and the solution in the form of the end of humanity were seen as something teleologically predetermined. Indeed, to a certain extent the problem was seen as identical with the solution (as we see in Strauß above) and the revival of the Schopenhauerian concept of the transcendence of human imperfectibility through the transcendence of humanity itself and a neo-Nietzschean interest in Buddhistic recurrence became common currency.

Today this idea has become so much part of the mainstream, if not quite so hysterical as in the early 1980s, that serious scientific researchers have turned to the idea that the future is already present and preordained and that it is to be welcomed. Jonathan Culler maintained as recently as 1988 that 'at the beginning of the eighteenth century (most) Protestants took the Bible to be the word of God. Now this is untenable in intellectual circles.'[26] I would contend that even just nine years later this now needs to be rewritten, in that although in intellectual circles the Bible as the word of God remains largely untenable, the question of existence *per se* and the problems associated with existence are increasingly approached in a neo-metaphysical way in both intellectual and non-intellectual circles. Even the bastions of science which were supposedly impervious to irrationalist and metaphysical thinking are beginning to reintroduce a teleological meta-narrative into the equation, with respected cosmologists such as Stephen Hawking and William Lane Craig positing that science is merely now a means by which we may push our way forwards to a rediscovery of God's plan, rather than the means by which we move away from a prejudiced and uncritical belief in the truth of such a plan. And certainly in the less scientifically rigorous circles of popular culture and ecological thinking, the idea of the rediscovery of a pantheistic spirituality of nature, the concept of an omniscient and omnipotent Gaia if not of the old paternalistic and monotheistic God of the Protestant faith, has made gigantic strides forward.[27]

But what is the connection with other contentions that the search for identity in nature is anti-democratic and anti-Enlightenment? In seeking to promote the needs of nature over those of people they have embraced an unfreedom – no democracy amongst a pack of wolves – which is precisely similar to that which they claim to escape from.

In its German form this trend has been particularly consistent, with a massive growth in support for neo-spirituality and the search for an individualistic identity which can replace that of nation, class, race or other collective entity, all of which seem to have been fundamentally discredited in the twentieth century. The postmodern condition, in its desire to reject objective truth claims, has created for itself a cosmic element to replace universalism. Postmodernism's retreat into individualistic positionality has recreated the loneliness and essential pessimism of solipsistic idealism which, in many postmodernists, has created the desire for an 'understanding' or, to give it its proper postmodernist sense, an 'intuition' of 'oneness', and that all things are connected. It could be argued that this form of cosmic postmodernist universalism differs in no way from previous 'modernist' forms of universalism in that it posits the existence of a totality and that, in any case, the modernist totalizing project was also cosmic in its utopianism.

However, the rediscovery of totality in the form of cosmic consciousness is in fact the precise reversal of the rational modernist Enlightenment project. The essential difference lies in the fact that the rational modernist project, above all in its Marxist form (which I believe is still the model which all critique is ranged against despite all the talk of the 'end of ideology'), posited that there was a totality of human existence, and that the point was both to comprehend and actively to change that totality in a way which would liberate humanity from the realm of necessity into the realm of freedom. Postmodernist cosmic universalism turns history back on its head and says that, yes, there is indeed a totality but that totality is *essentially* unknowable and has to be accepted at face value.[28] The nearest we can come to knowing totality is in submitting to a form of neo-Hegelian world spirit in the form of the Gaian spirit of the world. Where this becomes potentially anti-democratic is in the position adopted by Bahro at several points in his writing where 'Romantic individuals' step forward who claim to have a divine apprehension of this totality and the only true insight into the whole.

As a result, the 'holistic' approach moves to the centre of the ecocentric project, even though the use of this term is wholly inappropriate, representing as it does a purely individualized and atomized holism rather than a contextualized and socialized totality. The desire to treat the psychological and physiological problems facing the individual in the modern world by a retreat into holistic medicine and ideology is in effect nothing more than the ideological concomitant of social atomization. In attempting to treat the body as an isolated whole existing only in a natural environment, the holistic movement ignores the fact that the body exists in a social as well as a natural environment, and that many of the symptoms which drive them to holism are a reification precisely of the social atomization which holism reinforces.

It is possible to see in this question of ecological holism a clear parallel with Rorty's 'North Atlantic post-modern bourgeois-liberal neo-pragmatis(m)'. The question that has to be asked is where this trend has come from and where it could potentially lead. This apparent New Age of post-ideological pragmatism and 'end of history' triumphalism is not new at all but a restatement of the categories described by Nietzsche in *Ecce Homo*:

Wissen [knowledge]	*Natur* [nature]
Erkenntnis [cognition]	*Leben* [life]
Kritische Historie [critical history]	*Monumentale Historie* [monumental history]
Wissenschaft [science]	*Mythos* [myth]
Psychologie [psychology]	*Kunst* [art]
Gerechtigkeit [justice]	*Überzeugung* [conviction]
Das Wahre [truth]	*Das Schöne* [beauty]
(*Werden*) [becoming]	(*Sein*) [being]

In his positing of these alternative and, in his view, mutually exclusive categories, Nietzsche was, as we know, moving towards the concept of the superman, the person who would become what he is through the process of overcoming becoming and just being. In this process the transvaluation of all values would mean a rejection of all facets of truth, morality and so on. What Nietzsche and those who rest on his work, such as Bahro and most postmodernist thinkers today, seem unable to recognize is that there are two categories of truth and morality: (a) revealed or abstract truth, posited as teleologically predetermined and immutable, and (b) concrete or dialectical truth which is the materially determined dialogical reflection of social reality, historical change and progress and, above all, is based in social action rather than individualized passivity. It is the difference between passively accepted and actively constructed truth, the first of which is rightly rejected by those who accept the premises of the second.

The fundamental problem for postmodernism, post-structuralism and those who accept what Norris calls dogmatic relativism – and in the ecological context we could call species relativism, or biological egalitarianism[29] – is that while correctly resisting the Aristotelian concept of a revealed and immutable truth which is immediately apparent to the free-born citizen but had to be forced upon the slave, they go on to reject any attempt to arrive at an approximation of reality or truth which relies on logical and progressive argument in which it becomes necessary to convince people of a superior truth to that which they hold to be true. The arguments of those who resort to this form of dogmatic relativism are, paradoxically, even more coercive than those who argue for a hierarchy of ideas and the concept of false consciousness. For the concept of essentialist de-centred truth, i.e. that only gender, racial or species oppression rather than universalistic and class oppression are valid, leads them to a sort of hierarchical relativism which posits that only the person of correct gender, race or consciousness respectively can know the de-centred truth. At least with the old concept of truth it was available to all even if not by the same process. Now we can only look forward to the increased ghettoization of contingency.

This distinction between different sorts of truth can perhaps help us to understand precisely where postmodernism and radical ecology have their common roots, for they both issued, in their latest incarnation, out of a fundamental disillusionment with the modernist and Enlightenment traditions in their different forms. In attempting to escape the gravitational pull of universalistic concepts of truth, they have launched themselves into an orbit around the self which has brought them to an inherent remythologization and therefore remetaphysicalization of contingency, and represents the dialectical fusing of pre-modern romanticism and postmodern individualism. However, much of what purports to be subjective and 'resistant' reading is actually fully informed by a political agenda based on an appeal to truth, albeit a de-centred one. What is necessary is an investigation of the nature of the truths on offer, and whether they do actually contribute to the sum of human liberation.

It should not be forgotten that much of both the green and the postmodernist agenda stems from the atrophy of Marxist theory and praxis under the double impact of reformist social democracy and authoritarian Stalinism. However, the consequence of the collapse of the Marxist project, based in the concept of class struggle and the proletariat as the revolutionary subject with the objective capacity to liberate the whole of humanity, has not simply put the question of class on to the back burner but has resituated class as a threat *per se*. Nietzsche becomes central to this postmodernist trend because the basic impetus for his thought was a deep fear of the mass or herd which threatened the very existence, both physical and psychological, of the individual and individualism. For this reason, the disillusionment with the proletariat and the failure of the conscious proletarian movement in both east and west has fed through not simply into an *agnostic* attitude to the working class, seeing its failure not as a temporary setback, but into an essentially *antagonistic* attitude to the working class. Bahro himself gave vent to this point of view when he stated that

> the working class here [in Germany] is the richest lower class in the world. And if I look at the problem from the point of view of the whole of humanity, not just from that of Europe, then I must say that the metropolitan working class is the worst exploiting class in history.[30]

This complete inversion of traditional forms of rational analysis in which *aesthetic* and *ethical* judgements of the working class take over from rational and material ones, is the epitome of postmodernist thinking. The metropolitan working class in western Europe may well be the richest in the world, but this does not make it an exploiting class. On the contrary, it means that it is actually objectively the most exploited working class in the world in that the increase in labour productivity under western European capitalism has extracted so much surplus value from labour that the fruits of that expropriation could be relatively beneficially distributed thanks, not least, to the strength of reformist working-class movements in the west and the apparent threat of the eventual triumph of socialism in the east.

In any case, the age of a rich lower class in western Europe is on the wane. We are now well along the long downward wave of economic development which commenced in the early 1970s, and falling labour productivity and the reversal of the redistributive benefits of welfare capitalism have been issuing into renewed class conflict for some fifteen years.

But of course, the apparent retreat from class and ideology into individualistic positionality is itself an ideological response based in class interest. The postmodernist wave of individualism and the turn to cosmic ecologism above universalist humanism reflect the very real changes in socio-technological organization which have 'decentred' or 'deconstructed' production in the metropolitan centres, and which have brought with them a decline in 'Fordist' modes of production with large groups of working-class people centred on the

production of investment goods. The new flexibility, decentralization and intellectualization of production has fed into the growth of
decentralization, atomization and individualized intellectualization
as new 'ideologies'.

However, even the most committed ecocentric green has to
recognize that there exists a level of the problematic which cannot
be dealt with on the level of the individual. This is the potential
strength of the ecological movement as it restores the concept of an
objective problem existing outside the fantastic realms of individual
self-consciousness and the view that a solution to it must be found
on a 'universalistic' level. The major weakness in the eco-
movement is not this recognition but the 'deconstructed' political
conclusions which have been drawn from it.

In order to overcome this ecocentric shamanism it is necessary to
bring about a resocialization of progress and Enlightenment rather
than to abandon it. The Kantian antinomy between process and
change can equally be applied to nature, as nature is itself at one
and the same time a phenomenon and a process. It is undergoing
change all the time, with or without human intervention. Radical
ecology sees nature as an ahistorical phenomenon which should in
no circumstances be changed. This is explicitly nonsensical, as
nature and human beings only exist because of change and evolution. The phenomenon is the process, and green ideology, like all
bourgeois ideology, is static and ahistorical. It can see only its own
values as eternal 'common-sense' values, and at the moment we are
suffering a sort of eco-Fukuyamist 'end of history' in which the
post-structuralists have replaced God with the text, and the post-
materialist ecologists have replaced him with nature.

However, to quote Enzensberger again, 'it would be a mistake to
conclude that because of their boundless ignorance on social
matters, their (the ecologists') statements are absolutely un-
founded.'[31] How can ecology be rescued from the ecologists and
turned back into a central part of a wider social agenda?

Carl Friedrich von Weizsäcker maintains that 'humanity must win
the ability to reject what is technically possible if it fulfils no
purpose [my emphasis]. This requires self-discipline. Technology as
a cultural factor is impossible without the capacity for technological
asceticism.'[32] This, however, begs the question about who should
exercise the ascetic decision and why it is an ascetic decision rather
than a question of democratic control? The reason is that to use the

word 'control' would imply a set of structural rules based in an executive body carrying out political decisions, which all smacks too much of hierarchy for the postmodernist ecologist. Asceticism is, on the other hand, a nicely aesthetic and non-hierarchical word. However, the desire not to exercise control and take political decisions is an anarchistic hangover from the past. It is a tyranny of structurelessness of precisely the sort which can be exploited by the shamans and the 'princes'.

The problem is not too much democracy, as Bahro maintains, but too little. It is not man's inherent nature which is at fault but his distorted relationship with nature. The answer is not, however, to try and subordinate oneself to nature, to accept Aids, starvation, death and disease as Nature's way of telling us to slow down, but rather to re-establish democratic human control over our relationship with nature. The transcendence of nature over humanity will not liberate humanity but will only further enslave it. To return to a Schopenhauerian denial of causality in the name of Chaos and the free play of metaphor (Q: free from what? A: material reality) in the name of otherness is a reactionary step backwards rather than the radical step forwards often claimed.

The belief in nature as a God is a belief in a metaphysical 'salvation' and avoids the need for social action. Of course it is true that the future of humanity will only be free if it is based on a recognition of necessity, but only as the end of a process of human struggle rather than imposed by a benign bourgeois eco-dictatorship in the form of Bahro's theocracy.

In a letter to the present writer, Noam Chomsky says that he considers Hegel's *Philosophy of History* to be 'one of the most foolish and disgraceful books I have ever read'. I would go a long way in agreeing with this statement, but with one exception. In that work Hegel states that 'a people which considers nature to be its God can never be free.' This remains as true today as it was 150 years ago.[33]

Notes

[1.] David McLellan (ed.), *Karl Marx: Selected Writings* (Oxford, 1977), 64–74.

[2.] The concept of pragmatism should be defined here. The Green Party in Germany is now dominated by a pragmatic approach to political events which has placed members in positions of governmental responsibility and has essentially turned it into a left social democratic party with aspirations to collective action. However, the term as used below in its Rortian sense refers to a brand of neo-pragmatic individualist philosophy as expounded by the American liberal philosopher John Dewey in the 1930s. The two should not be confused.

[3.] See, for example, Statham in ch.7 of this volume.

[4.] His books include *From Red to Green* (London, 1984); *Radikalität im Heiligenschein. Zur Wiederentdeckung der Spiritualität in der modernen Gesellschaft* [*The Radical Halo. On the Rediscovery of Spirituality in Modern Society*] (Berlin, 1984); *Logik der Rettung. Wer kann die Apokalypse aufhalten?* [*The Logic of Deliverance. Who can Stop the Apocalypse?*] (Stuttgart and Vienna, 1989).

[5.] *Der Spiegel*, 26 (1992), 62–3.

[6.] For example, some with apparently unimpeachable democratic credentials such as Jens Reich have argued in a similar vein. See the Jens Reich interview in *Der Spiegel*, 14 (1995), 42–9 where he states that 'real change is not possible as long as constant elections are getting in the way.'

[7.] See note 4 above.

[8.] Bahro, *Logik der Rettung*, 431.

[9.] Richard Rorty, 'Moral identity and private autonomy: the case of Foucault', in *Essays on Heidegger and Others* (Cambridge, 1991), 194.

[10.] *Sämtliche Werke*, 1, 164 quoted in Lukács, 'The bourgeois irrationalism of Schopenhauer's metaphysics' in Michael Fox (ed.), *Schopenhauer: His Philosophical Achievement* (Brighton and Totowa, 1980), 183–93.

[11.] Ibid., 165.

[12.] Bahro, *Logik der Rettung*, 434.

[13.] Hans Magnus Enzensberger, *Raids and Reconstructions* (London, 1976), 262 (English translation of *Palaver. Politische Überlegungen (1967–1973)* (Frankfurt am Main, 1974), 180). See also *Kursbuch 30* (1972) and 'Ökologie und Politik oder Die Zukunft der Industrialisierung', *Kursbuch 33* (1973), 1–42.

[14.] Hans-Georg Betz, 'The post-modern challenge: from Marx to Nietzsche in the West German alternative and green movement', *History of European Ideas*, 11 (1989), 815–30, here 816.

[15.] Herman Hesse, *Der Steppenwolf* (Frankfurt am Main, 1970), 59.

[16.] Peter Thompson, 'Progress, reason and the end of history', *History of*

European Ideas, 16 (1994), 361–71.

[17.] Bahro, *Logik der Rettung*, 432.

[18.] Ibid., 439.

[19.] Enzensberger, *Raids and Reconstructions*, 274.

[20.] Ibid., 293.

[21.] David Pepper, *Eco-socialism: From Deep Ecology to Social Justice* (London and New York, 1993), 147.

[22.] Ibid., xvii.

[23.] Ibid., 148.

[24.] Botho Strauß, *Trilogie des Wiedersehens/Groß und klein: Zwei Theaterstücke* (Munich, 1980), 138–9.

[25.] *Jugendwerk der Deutschen Shell: Jugend 81*, 1 (Hamburg, 1981), 378.

[26.] Jonathan Culler, *Framing the Sign: Criticism and its Institutions* (Oxford, 1988), 79.

[27.] Also see Frank J. Tipler's book *The Physics of Immortality*, which was at the top of the best-seller list in Germany for fifteen weeks and which posits that at some point in the future God will be revealed as an infinitely powerful computer in which all life and all potential life will be instantly recreated. This is a prime example of the recent trend towards the pseudo-scientification of metaphysics.

[28.] Jürgen Habermas, *Nachmetaphysisches Denken: Philosophische Aufsätze* (Frankfurt am Main, 1992), 35.

[29.] Pepper, *Eco-socialism*, 3.

[30.] Bahro, *From Red to Green*, 184.

[31.] Enzensberger, *Raids and Reconstructions*, 290.

[32.] Quoted in Peter Weinbrenner, *Die Zukunft der Industriegesellschaft im Spannungsfeld von Fortschritt und Risiko* (Bielefeld, 1989), 20.

[33.] G. W. F. Hegel, *Philosophie der Geschichte* (Stuttgart, 1961), 145 (my translation).

7 • Ecology and the German Right

Alison Statham

Discussing ecology as an element of right-wing thought might appear misguided given the political direction of Die Grünen, today's most significant representative of environmental ideas in Germany. Yet while the left-liberal wing of the green movement certainly holds centre stage, authoritarian models of anti-modernist, anti-technological and even ecological thought do exist. Dirk Fleck's 1993 novel *GO!*, for example, predicts the establishment of a future 'Ökodiktatur' [eco-dictatorship] if society fails to act urgently against environmental destruction.[1] Right-wing solutions to ecological problems are sought in the pages of ultra-conservative, if not extremist journals such as *Nation Europa*, *Criticón* and *Mut*. Moreover, the modern green movement encompasses a number of conservative ecological parties engendered by co-founders of Die Grünen who opposed that party's swing to the left at the Saarbrücken conference of 1980. The principal parties to consider are the Ökologisch-Demokratische Partei [Ecological-Democratic Party], established in 1981, and the Unabhängige Ökologen Deutschlands [Independent Ecologists of Germany], established in 1991. Both were formed under the influence of the erstwhile *Union* member, Herbert Gruhl, the latter party evolving from elements accusing the former of gravitating too closely to the *Union* [Conservative] parties. They also, however, possess less immediate antecedents, for an examination of ultra-conservative traditions reveals a historical continuity of ideology.

The destructive effects of industrialization upon rural idylls, as well as its modernizing consequences on all aspects of life, have formed a fundamental strand of conservative and ultra-conservative thought in Germany for over a century. Its all-embracing character means that anti-modernism has attacked not only industrial and technological advance, but also the sweeping social and political changes brought about in its wake, namely democratization and egalitarianism.

The organicism which developed during the Napoleonic occupation and was later expanded as a counter to greater working-class influence, female emancipation and racial equality during the late nineteenth century, lives on today, albeit in an adapted and to some extent diluted form which owes something to its historical burden. Indeed, the switch to the use of ecologism as a definition rather than the former term, 'organicism', constitutes an attempt to move away from discredited historical forerunners. Nevertheless, 'ecology' would not appear to be a wholly inappropriate label, given the extent of overlap between targets of attack for right- and left-wing opponents of excessive technological advance.

The convergence of left and right attitudes towards the consequences of modernization is no post-war phenomenon. On the contrary, both poles of opinion not only have an established tradition of concern about urban deprivation, but they also attack materialism. Commonality of purpose may not always occur. For example, the worker literature and poetry of the early twentieth century[2] focuses upon the materialism of producers who live off the fruits of proletarian labour, whilst the middle class and its representation utilized the *völkisch* ideology of natural hierarchy to assert that Social Democracy was materialist in encouraging workers to demand more than was 'naturally' theirs. Many of the late nineteenth-century pressure groups, largely middle- and upper-class in composition and attitude, became some of the staunchest advocates of *völkisch* ideas, if for nothing else, then to defend their own selfish interests against the perceived threat of impending proletarianization.[3] The works of leading conservative ideologues provided more than sufficient fuel for the forces of political reaction. Basing the success of their ideal nation upon obedience to so-called laws of nature, the likes of Paul de Lagarde and Julius Langbehn were able to lay the blame for Germany's alleged decline on un-German attitudes and political forms advocated by those striving for greater social and racial equality. In the words of Langbehn, 'equality is death, hierarchy is life'.[4]

Nationalist *völkisch* ideology provides the basic concept for modern right-wing ecological thought and practice. The idea of being rooted in a particular *Gemeinschaft* [community] and landscape by shared language, history, traditions and ancestry is fundamental to the right's recognition of *Umweltschutz* [protecting the environment] being synonymous with *Lebensschutz* [protecting life], *Heimatschutz* [protecting the homeland] and *Volksschutz* [protecting the people/nation]. Whereas the pre-1933 and Nazi Heimatschützer wrote explicitly of bacteria and disease laying the German *Volk* and its culture waste, the rhetoric of post-Nazi intellectuals looks to the concept of cross-cultural pollution and loss of identity.[5]

It is within the realm of urban deprivation versus rural idyll that the strongest evidence of conservative involvement in early green thought becomes clear. Industrialization and urbanization are charged with destroying the life-blood of the *Volk*, the *Bauerntum* [peasantry]. As the *Bauerntum* is being depleted, so the *Volk* itself is driven from its *Gemeinschaft* to lead a life of materialist individualism in the town, the 'stone desert, which dehumanizes the *Volk*',[6] and which houses the ultimate antithesis of organic living, the Jew. With this loss of *Gemeinschaft* and therefore 'all-embracing unity',[7] the *Volk* becomes a 'non-organic mass',[8] open to further destruction as it is now sufficiently degenerated to receive democracy and liberalism.

Objections to urban life were based on aesthetic taste rather than scientific reason.[9] Nevertheless, attempts to recreate communities in rural settings are of interest, not only because of their often racial emphasis, later to be adopted and 'refined' by the Nazis. They are, racial policies apart, the historical predecessors of those experiments in communal lifestyles practised during the sixties and seventies. Such colonies, for example Eden (1890s) and the 'Artamanenschaften' (Weimar), emphasized the importance of communal, as well as national, self-sufficiency and shared obligation to the community.[10]

In the post-war period, conservative ecologism and nationalism found considerable scope for development within the anti-nuclear and peace movements. For a lobby determined to campaign for German unification and freedom from allied control, demonstrations against the installation of nuclear warheads and power stations, and for German bloc independence, offered ample opportunity to liken the peace movement to a force fighting for German self-liberation from foreign rule. The journal *Wir Selbst* was a bastion of this theme throughout the eighties, and illustrated with great effect the extent of convergence between right- and left-wing ideology by commenting on and publishing literature representing both viewpoints. The mixture of right nationalism with left internationalism, as well as the conscious use of left-alternative rhetoric, was intended to benefit an ethnopluralism which rejected standard western and eastern models.[11] The journal even manages to describe itself and its contributors as 'socialist'. Paul Frister writes on a subject familiar to the readers of *Wir Selbst* in distinguishing between socialism, which supports the 'supremacy of the people', and Marxism, advocating a materialist class society. Socialism is neither capitalist nor communist. Instead it seeks to create an independent state engaged in the 'German people's struggle for national freedom' against the dominance of the superpowers.[12] Such arguments illustrate a significant continuity of theory with that of the Weimar 'conservative revolution', most notably with the ideas of Thomas Mann and Moeller van den Bruck.[13]

A further element of overlap is apparent in the calls for greater democracy from ecological conservative sections. Again it is within the pages of *Wir Selbst* that appeals for *Basisdemokratie* [grassroots democracy] may be found. It goes without saying that this is still a relevant theme for Die Grünen today. As it was prevalent amongst those organizations and initiatives which co-founded the party, the principle of participatory democracy lives on in the conservative groupings that left in protest at the post-Saarbrücken left-wing emphasis. The organ of the Unabhängige Ökologen Deutschlands (UÖD), *Ökologie*, regularly publishes articles which call for Germany and Europe to adopt a federalist structure whilst attacking alleged impulses towards greater centralization of power as dictatorial.[14] Consequently, it is not only parties supporting national independence from European bureaucracy that are praised by *Ökologie*. Those campaigning for the protection of regional diversity within Germany, for example Hubert Dorn's Bayernpartei [Bavarian Party], also receive considerable coverage. Indeed, it is perhaps for the protection of national and regional identity from levelling influences, rather than out of any real concern for democratic principles, that *Basisdemokratie* is advocated within this section of the right.

The preference for regionally based decision-making brings conservative ecologists to the concept of bio-regionalism. As the regional model is perceived as the best means to establish a *Basisdemokratie*, the creation of bio-regions is seen as the way forward if global environmental disaster is to be averted. By utilizing geographical and cultural uniqueness as a basis for political action, the UÖD and other European conservative ecologists, for example, France's Mouvement Écologiste Indépendent, propose to develop an ecological, federalist alternative to the EU away from a course of economic concentration and political centralization. Autonomy for Scots, South Tyroleans etc., to oversee the use of their own natural resources would do more than preserve their unique identities. The UÖD argues that this would provide their best defence against 'the plundering of resources, garbage tourism and economic levelling'.[15]

Ecological discourse, used by ultra-conservatives to cover a wide cross-section of themes, thus forms a major constituent of right-wing thought. While right-orientated ecologists may shy away from the blatant racist rhetoric of the neo-nazis, the underlying motif of arguments for the mutual protection of cultural and national identities and for the retention of ethnic diversity can be substantiated by the well-known skinhead battle-cry, 'Ausländer raus!' [Foreigners out!]. Even those environmental concerns in common with the left are linked to racism. For example, the *Gastarbeiter* in Germany are condemned for plundering already overburdened resources (see below).

Despite the clear convergence of themes, the conservative lobby has been quick to accuse the left, and particularly Die Grünen, of being Marxist and therefore too limited in their approach by allotting ecology a political label. Herbert Gruhl's speech at the 1982 ÖDP conference begins to explain the conservative definition of ecology as about life itself, rather than party political manoeuvring.[16] The issue is discussed further in numerous issues of *Ökologie* whose contributors seek a solution to old political division and describe conservative ecological policy as 'neither right nor left'.[17] Such an assertion brings to mind the Weimar Right's pursuit of a third way between Marxism and capitalism. Exactly this is attempted by Germany's self-professed *Wertkonservativen* [value conservatives], who emulate the ideas of their inter-war predecessors. Attacks on capitalism and Marxism as equally materialist have been frequent throughout the post-war era. Anton Büchting of the Deutscher Block [German Block] illustrates the conservative stance perfectly: 'In the west individual exploitation, in the east collective exploitation ... essentially, capitalism and bolshevism are the same.'[18]

More recent criticism of left-wing ecologism has been much sharper. Die Grünen are accused of turning ecology into a 'class struggle slogan' far removed from the sense of community and obligation towards fellow human beings which harmonizes environmental protection with 'the laws of life'.[19] Even more extreme is the claim that Green Party activists, as Marxists, are as materialist as the system they are supposed to be condemning. A movement which approves of squatting, abortion and multiculturalism has, in the words of Peter Dehoust, 'understood nothing of ecology and the protection of life in the broadest terms'.[20]

Such allegations are, of course, intended to extend credibility to conservative ecologism in an age when arguments for obedience to the laws of nature are heavily burdened. The assumption that this theory of environmental protection is tempered by no specific political inclination is clearly false. The view of liberalism as the ultimate antithesis of life in accordance with nature is typical of more extreme strands of right-wing ideology. It is liberalism which has facilitated the development and acceptance of those elements constituting unecological living. In that it encourages immigration and perceived materialist lifestyles, liberalism, the arch-enemy of the Right, is seen to be the root of Germany's 'environmental' problems.

The immigration and integration of foreigners into Germany is the main issue for conservative ecologists. Coexistence of traditions, cultures and values which are irreconcilably different is understood to be non-organic, as the multiculturalism it promotes interferes with natural variety. Wolfgang Seeger offers an explicit explanation of the right's attitude by suggesting nature created neither a 'standard animal', a 'standard plant', nor a 'standard person'. Instead the key to survival is the 'variety and diversity of the different species and races'.[21] The net result of attempts to defy this natural diversity is usually depicted in apocalyptic terms. Cultural pollution will not only destroy traditional values for immigrants and native populations alike. The very future of humanity is deemed to be at stake. Less alarmist, if uncompromising, visions predict the creation of a 'McDonalds–Drogen–Sklaven–Einheitsbrei–Unkultur' ['McDonalds–drugs–slaves–standard brew–non-culture'].[22]

Immigration is not the only enemy where foreign values are perceived to be penetrating and despoiling German culture. The European Union has come under fierce attack from the German Right throughout its existence. As early as 1957 the Sozialorganische Ordnungsbewegung Europas (SORBE) [Movement for

Social and Organic Order in Europe], whilst acknowledging Europe's obligatory alliance against US and Soviet domination, campaigned for the preservation of the biological substance of all peoples and the retention of national independence and identity.[23] The emphasis of a European community was, according to SORBE and other representatives of the National Opposition, to be above nationalism, rather than seeking 'a vague internationality, or even a deracinated, mixed-up human soup'.[24] Various crises during the EU's history have been presented as examples of its obvious ultimate failure. Mrs Thatcher's position was supported in the eighties, her determination to reduce Britain's budget contributions being hailed as a victory for national consciousness.[25] The budget crisis was portrayed also as conclusive proof that the diversity of national interests between member states was too great for the then Community to survive whilst it persisted with pursuing 'supranational fantasies'.[26]

Whilst the infiltration of alien cultures and values into Germany through both immigration and European integration is regarded as making Germans 'strangers on their own soil',[27] the right also emphasizes the negative effects of immigration and multiculturalism on the foreigners as equally catastrophic. While Germany itself may be facing an excess of foreigners, the representation of individuals divorced from their 'cultural and national individuality'[28] is an important feature of post-Nazi anti-immigration theory. The term *Entwurzelung* [deracination], used of the threat to Germans from various facets of modern life at the start of the twentieth century, has re-entered the right's vocabulary. But this time it is also being used to illustrate the situation facing those children born in Germany to immigrant families. These children are seen to belong nowhere. Born into an alien culture and isolated from their true traditions, they are without a *Heimat*, the most fundamental form of national and racial identity. By defending every individual's right to a *Heimat*, the ecological right seeks to legitimize its essentially racist ideology.[29] The argument prevails that this lack of true identity will cause young foreigners in Germany to feel resentment towards their plight. Forced to reconcile themselves to wholly unecological circumstances, they are likely to turn to criminal activity as a means of expressing their discontent. In addition, pessimistic visions predict an escalation in racial tension as young Germans and non-Germans alike seek to discover who they really are[30] in a future when multiculturalism will strive to redefine the criteria for membership of a nation or *Volk*.

The right-wing emphasis on *Heimat* brings conservative thinkers to their policy on the extension of German nationality law. The basic argument here is that citizenship alone does not make a German. Racial ancestry means that those born of non-German parents can never be German themselves, since they are incapable of feeling German. Wolfgang Seeger sums up this stance as follows: 'you can change your state allegiance many times, but never your nationality!'[31] The only solution to this irreparable discrepancy between place of residence and *Heimat* is envisaged as repatriation assisted by a programme of financial and economic aid to be funded by the federal government and with the intention of making immigration unnecessary and undesirable.[32]

Regardless of whether the post-1945 depiction of foreigners as equal victims of materially driven immigration in certain circles is based on a genuine concern for their welfare, it has not been without impact. Non-Germans may not have flocked to join those parties and groups opposed to their presence, but a few *Gastarbeiter* now in Germany for a considerable time have taken seriously the claims of more moderate groups to defend their interests. The key here is the distinction between *Gastarbeiter* who are seen to have made valid contributions and commitments to Germany, and the so-called *Scheinasylanten* [fraudulent asylum-seekers] who are accused of entering the country for personal economic gain under the guise of being politically persecuted. A notable example of established *Gastarbeiter* interest in this alleged consideration has been provided by Hacer Rieb, the Turkish-born wife of a Hanover police official and strong opponent of the asylum laws prior to their 1993 revision.[33]

Nevertheless, this historically burdened approach is by no means universal within the right. Today's Judaism is Islam and certain quarters attack it as viciously. The availability of immigration in the first place is held responsible for the inclination of yet more foreigners to come to Germany. That there is a channel by which to escape the poverty of the Third World is perceived, for example, by the Schutzbund für das Deutsche Volk (SDV) [Association for the Protection of the German People] as creating an unwillingness within immigrants' countries of origin to reform and improve conditions. This then leads to further immigration as even more people attempt to flee the essentially restrictive life of Islam.[34] Simplistic solutions offered by the SDV in particular are of little help. Arguing that population control in Islamic nations would reduce the need to emigrate fails to appreciate the right's own message: that Christian and Islamic cultures differ fundamentally. The idea of educating Islamic nations about family planning not only contradicts the ecological right's alleged respect for variety. The idea also directly contravenes the movement's own abhorrence of widely available family-planning services which are deemed to be threatening the Germans with extinction.

More sinister than the accusations of the pursuit of irresponsible domestic policies within Islamic nations, is the perception of claimed hegemonic tendencies. Hubert Dröscher (SDV) hints at immigration as an instrument of 'power-political forces' working towards the 'distribution of Islam through Europe'.[35] By comparison with the way in which conservative revolutionaries assessed the Jewish presence before the Nazi take-over, so-called Islamic intent is interpreted on the whole more moderately by the non-Nazi elements within the right. Nevertheless, accusations that 'Islamists are on the ascendancy in Germany'[36] have factors in common with the anti-Semitism of Weimar writers. Rosenberg's assertion that Jews' involvement in all aspects of modern life was part of a plot to destroy Europe and establish a 'Jewish tyranny'[37] is not so far removed from modern interpretations of Islam.

Allied re-education is also made to bear responsibility for the increased immigration which allegedly threatens Germans with extinction. Not only is the programme considered to have been imposed by the occupying forces who divided the country, forced it to join the European Community and denied it full sovereignty, but re-education also, the argument runs, treated Germany as wholly to blame for the start of the Second World War, thus hindering the development of a 'healthy feeling of national self-worth', an impediment which, in turn, reduced the German 'will to reproduce'.[38] According to the right, this programme renews the Allied plot of the First World War to overpower and strip Germany of its unique identity and role as a balance between east and west, and to force the nation to conform to the European norm.[39] In other words, it constitutes 'a continuation of the war by other means'.[40]

The occupation of a divided Germany may be depicted as unnatural by the right, and may have led to its involvement in the peace movement alongside left-wing factions. However, the description of Bonn and other governments as Reagan's 'unconditional vassals' comes from Petra Kelly.[41] The left, as well as right-wing ideologues, frequently lamented a West Germany living under 'American supreme command',[42] whose political role was no more than that of 'lackey of the foreign rulers'.[43] Indeed, a strong element of convergence in rhetoric between right and left offers another insight into how ideologies have become entangled in the post-war era.

Materialism is another key factor when considering responsibility for immigration. An age-old enemy of the ultra-conservatives, its force and appeal continues to disturb those figures who still regard it as destroying traditional German values. As far as industrial expansion and consumerism are concerned, left and right agree on their debilitating effect on natural resources. However, a uniquely right-wing form is evident in that the need for *Gastarbeiter* is attributed to the greed of industrialists and consumers. In addition, those Germans refusing to perform menial tasks bear the blame for the influx of foreign labour. If Germans had not been so materialist in the fifties, it is argued, cultural pollution from immigration would not have occurred. Axel Schnorbus asserts: 'Germans too can pick cherries, carry suitcases, fill cans of fish or work on an assembly line.'[44]

Unecological attitudes are not solely associated with immigration and multiculturalism. Antisocial behaviour is linked to the collapse of the germ plasm of the *Volk*, the family, due to materialism. Although female emancipation is not condemned as evidence of racial degeneracy as it was by the inter-war Deutschvölkischer Schutz- und Trutzbund [German People's Offensive and Defensive Association],[45] the post-war right does hold the career woman responsible for the decline of the family in that she is disobeying the laws of nature. Close links to historical forerunners can be detected here. Edgar J. Jung leads the pre-Nazi attack by condemning feminists for destroying *Mütterlichkeit* [motherliness].[46]

The connection between emancipation and degeneracy is not broken completely by modern thinkers. Now the collapse of the traditional family due to the changing expectations of women is associated with juvenile delinquency. The Grüne Aktion Zukunft (GAZ) [Green Future Action], claims to have proof that a child's later development will be damaged if the same person is not responsible for its upbringing during its most formative years. The assertion that 'it is in early childhood that the foundation stones are laid not only for most subsequent behavioural problems of young people, but also for neuroses and psychoses in later years', lays the blame for youth crime on those mothers who would allegedly rather help raise the GDP than responsible human beings.[47]

The proposed solution to this crisis is interesting in that it appeals to the base materialism deemed to be its cause. The payment of a 'skilled worker's wage' to mothers of four or more children would be funded by the deportation of immigrants.[48] Such a proposal is as simplistic as the argument that mass immigration causes a drop in the German birth rate, thereby leading to further immigration to make up the shortfall.[49] However, simple solutions to complex problems are a speciality of the right.

Regardless of their moral repugnance, authoritarian versions of green thought are consistent with the traditional and fundamental values of ultra-conservative opinion. That the less offensive aspects of conservative ecology are those very areas where convergence with the left alternative movement is most clear, does not necessarily imply that the right is hijacking more credible ideals for its own advancement. These areas, too, reveal considerable scope for the continuation of deeply conservative principles. The anti-nuclear issue provided an opportunity to campaign for reunification and the removal of the foreign domination of German affairs. The call for greater democracy is intended less as a means of achieving moral credibility than as an instrument of the right's anti-internationalist approach. Both themes have an established position within right-wing ideology.

Anti-modernism too is historically a facet of conservative thought. In that it attacks the encouragement of materialist urges and impulses which upset traditional practices and perceptions, the anti-modernist theme is more at home with conservativivism than the left progressivism of Die Grünen. Indeed, leading ideologues argue that no change in ecological circumstances can be expected from Die Grünen. Bahro explicitly suggests that the necessary transformation can only occur with the '*momentum* of a conservative revolution'.[50] That Hitler has given Germany a place in history as 'the land of the conservative revolution',[51] should not deter Germans from this course, Bahro feels. However, it should not be assumed that another Hitler is considered a desirable means to an ecological end. Rather Bahro appeals for an *Über-Hitler* [Super-Hitler] to fight against the 'atomization of society' and protect human substance.[52] Who this *Über-Hitler* should be and what powers he would wield is not clear. What is clear, however, is Bahro's belief that the conservative lobby holds the key to ecological renewal, and not Die Grünen.

This stance is in line with the origins of conservative political ecology which began as an anti-democratic movement during the nineteenth century. It is false, therefore, to assume that ecological impulses are part of a post-war and specifically left-wing phenomenon. On the contrary, they have a long tradition in German and western thought and have far less reputable origins than might be assumed. Anna Bramwell assesses this situation clearly: 'Greens may be ecologists, but not all ecologists are Greens.'[53]

Notes

[1.] See Heinz-Siegfried Strelow, 'Kultfrau Xenia. "GO!": Dirk Flecks Roman über Ökodiktatur', *Ökologie*, 6, 4 (1994), 20–1.

[2.] See for example Jean Baptist von Schweitzer, *Ein Schlingel. Soziales Bild in einem Akt* (1867).

[3.] A comprehensive study of Germany's pre-1914 pressure groups is provided by Geoff Eley, *Reshaping the German Right: Radical Nationalism and Political Change after Bismarck* (New Haven and London, 1989).

[4.] Julius Langbehn, *Rembrandt als Erzieher* (Leipzig, 1890), 153.

[5.] Bernhard Barkholdt, 'Die Bürgerinitiative Ausländerstopp', *Nation Europa*, 31, 2 (1981), 33–8.

[6.] Edgar J. Jung, *Die Herrschaft der Minderwertigen. Ihr Zerfall und ihre Ablösung durch ein neues Reich*, 3rd edn. (Berlin, 1930), 163.

[7.] Oswald Spengler, *Untergang des Abendlandes*, 1 (Munich, 1928), 34.

[8.] Ibid., 449.

[9.] See Matthew Jefferies on Paul Schultze-Naumburg in ch.2 of this volume.

[10.] A study of pre-Nazi communes is offered by Ulrich Linse (ed.), *Zurück, o Mensch, zur Mutter Erde. Landkommunen in Deutschland, 1890–1933* (Munich, 1983).

[11.] Armin Pfahl-Traughber, 'Brücken zwischen Rechtsextremismus und Konservativismus. Zur Erosion der Abgrenzung auf publizistischer Ebene in den achtziger und neunziger Jahren', in Wolfgang Kowalsky and Wolfgang Schroeder (eds.), *Rechtsextremismus – Begriffe, Methode, Analyse* (Opladen, 1994), 160–82.

[12.] Paul Frister, 'Sozialismusdiskussion', *Wir Selbst*, 2 (1980), 11–13.

[13.] Thomas Mann, *Betrachtungen eines Unpolitischen* (Berlin, 1919); Arthur Moeller van den Bruck, *Das Dritte Reich*, 3rd edn. (Hamburg, 1931).

[14.] Kristof Berking, 'Die Neugliederung Deutschlands', *Ökologie*, 7, 1 (1995), 13–18.

[15.] Michael de Wet, 'Neue Götter braucht die Welt', *Junge Freiheit*, 25 August 1995.

[16.] Herbert Gruhl, 'Ökologische Politik zum Überleben notwendig', *Nation Europa*, 32, 5 (1982), 25–35.

[17.] Antoine Waechter, 'Weder rechts noch links!', *Ökologie*, 6, 1 (1994), 4.

[18.] Anton Büchting in *Die Brücke. Berichte über die Nationale Opposition*, 11, 7 (1955), 10–11.

[19.] Rolf Kosiek, 'Geistige Grundlagen der Umweltzerstörung', *Nation Europa*, 32, 5 (1982), 5–10.

[20.] Peter Dehoust, 'Zu diesem Heft', *Nation Europa*, 32, 5 (1982), 3–4.

[21.] Wolfgang Seeger, *Ausländerintegration ist Völkermord. Das Verbrechen an den ausländischen Volksgruppen und am deutschen Volk* (Pähl, 1981).

[22.] 'Was geht hier eigentlich vor?' *Unabhängige Nachrichten. Nachrichtendienst und Mitteilungsblatt unabhängiger Freundes-Kreise*, 12 (1994), 3.

[23.] 'Die Sozialorganische Ordnungsbewegung Europas', *Die Brücke, Auslandsdienst*, 4, 8 (1957), 2–6.

[24.] Max Sesselmann, 'Die sozialorganische Wirtschaft', *Die Brücke*, 4, 20 (1957), 2–3.

[25.] Heinz Petry, 'Der große Garaus von Brüssel', *Deutsche Monatshefte*, 35, 4 (1984), 3–4.

[26.] Heinz Petry, 'Die wirtschaftlichen Schäden der EG-Sucht', *Deutsche Monatshefte*, 35, 5 (1984), 5–6.

[27.] Karl Richter, 'Alle Menschen sind Inländer – daheim', *Nation Europa*, 44, 6 (1994), 3–5.

[28.] Barkholdt, 'Die Bürgerinitiative Ausländerstopp', 34.

[29.] See Jürgen Hatzenbichler, 'Heimatrecht – Recht auf Heimat', *Wir Selbst*, 1 (1995), 16–20.

[30.] Manfred Müller, 'Gastarbeiterkinder: Die Zeitbombe tickt', *Nation Europa*, 28, 12 (1978), 27–31.

[31.] Wolfgang Seeger, *Der Untergang der Völker Europas in einem eurasisch-negroiden Völkergemisch* (Tübingen, 1984), 4.

[32.] Helmut Schröcke, 'In die Heimat helfen', *Nation Europa*, 31, 12 (1981), 37.

[33.] Michael Haller and Gerhard C. Deiters, 'Alte Parolen, neuen Parteien', in Wolfgang Benz (ed.), *Rechtsextremismus in der Bundesrepublik:*

Voraussetzungen, Zusammenhänge, Wirkungen (Frankfurt am Main, 1989), 248–72.

34. From a letter dated 25 January 1995 which accompanied a variety of pamphlets published by the SDV following a request for further information on the group's ecological viewpoint.

35. Ibid.

36. Wolfgang Steinmann, 'Allah Pur', *Europa Vorn*, Spezial No.9 (1995), 12–13.

37. Alfred Rosenberg, *Die Protokolle der Weisen von Zion und die jüdische Weltpolitik*, 4th edn. (Munich, 1933), 44.

38. Schutzbund für das Deutsche Volk, 'Die gesteuerte Überfremdung', campaign leaflet no. 9405-30-2.

39. See Thomas Mann, 'Gedanken im Kriege', in *Essays*, 1 (Frankfurt am Main, 1977), 188–205.

40. Heinz Petry, 'Der große Garaus von Brüssel', 3.

41. Introduction to Gert Bastian, *Notwendige Anmerkungen zum NATO-Doppelbeschluß in der Darstellung der Bundesregierung* (Bonn, 1983), 3.

42. Hans-Dietrich Sander, 'Die Raketen stehen', *Deutsche Monatshefte*, 35, 1 (1984), 3–4.

43. August Haußleiter, 'Der Ausverkauf Deutschlands', *Deutsche Gemeinschaft*, 15, 4 (25 January 1964), 1.

44. Cited by Manfred Müller, 'Gastarbeiterkinder', 31.

45. Deutschvölkischer Schutz- und Trutzbund Ortsgruppe Meissen (ed.), *Eine unbewußte Blutschande. Der Untergang Deutschlands. Naturgesetze über die Rassenlehre* (Meissen, no date), 15.

46. Jung, *Die Herrschaft der Minderwertigen*, 195.

47. Grüne Aktion Zukunft, *Das Grüne Manifest. Programm der Partei 'Grüne Aktion Zukunft'* (1978).

48. Karl Baßler, 'Was die Rechte braucht', *Deutsche Monatshefte*, 38, 5 (1987), 6–10.

49. Hubert Dröscher, *Das deutsche Volk in der Todesspirale. Bevölkerungsentwicklung und Bevölkerungspolitik in der Bundesrepublik. Systembetrachtung und Gegenvorschlag* (Coburg, 1987), 9–12.

50. '"Eine ökologische Wende ist unmöglich ohne das Moment einer konservativen Revolution". Rudolf Bahro über die politischen Vorstellungen Kurt Biedenkopfs', extract from *Logik der Rettung. Wer kann die Apokalypse aufhalten*, in *Ökologie*, 7, 3 (1995), 13–14.

51. Ibid., 14.

52. Ibid., 14.

53. Anna Bramwell, *Ecology in the Twentieth Century: A History* (New Haven and London, 1989), 20.

8 • Ecofeminism in a Divided Germany: Irmtraud Morgner's *Amanda*

Antje Ricken

What have I seen? . . . I have seen frightening things. For example: the four great biological systems on which humans depend . . . are already overloaded. Resources indispensable for the survival and sustained development of humankind are being destroyed or exhausted ever more quickly; the demand for these resources is still growing. If the deterioration of the soil continues at the present rate, a third of the earth's arable land will become unusable over the next twenty years. Within this time, the world population will in all likelihood increase by half again. An ever-growing number of human beings thus requires resources which are becoming ever scarcer.[1]

So begins one of the reports on the state of the world which the snake Arke, a daughter of Gaia, brings back from one of her world explorations for the information of the siren Beatriz – fictitious narrator of Irmtraud Morgner's novel *Amanda* (1984). Almost like a flying Greenpeace monitoring vessel, the messenger Arke gives an account of the destruction of flora and fauna, the pollution of the oceans, the cutting down of rain forests and the increasing nuclear threat – these are enumerations with which we have become all too familiar over the last decade or so through the western media. During the late seventies and early eighties in West Germany, the issue of the environmental threat not only gained hold of the public consciousness, but also entered politics with the founding of the Green Party in 1979. It was during this time that Irmtraud Morgner's novel *Amanda* was written, adding a distinctively East German voice to the chorus of concern.

My students in Swansea, who grew up in the eighties, have been confronted from an early age with the litanies of environmental horrors and are well used to them. 'The environmental crisis' proves one of the most popular topics for oral presentations. With great fluency they list the apocalyptic facts of the global threat, but only to conclude that we must recycle glass bottles, buy aerosol-free

sprays and leave the car in the drive more often. A regretful shrug
of the shoulders signifies that environmental destruction is considered inevitable in a modern society.

It is against such assumptions of inevitability, against the
paralysing effect of apocalyptic prospects and against a view of the
environmental crisis that focuses only on symptoms that Morgner
launches an imaginative inquiry into the causes which may have led
to the present ecological imbalance. The approach she chooses is
that of a feminist critique of western civilization: 'Conqueror-thinking in society, science and technology has driven the earth to the
brink of abysses. Conqueror-thinking of men – culturally cultivated,
not naturally male' (*A*, 306).

Patriarchal imperialistic thinking as root cause of the environmental crisis – this suggestion places Morgner's approach close to
that of ecofeminism, which emerged from the radical ecology
movement during the eighties, principally in Britain and the USA. I
should like to show in this chapter how Morgner in her novel translates this kind of radical critique of civilization into literature, and
further, how – working in and from the context of GDR literature –
she anticipates the central issues of the ecofeminist debate.

Ecofeminism became an identifiable movement in the USA in
1980 with a major conference on 'Women and Life on Earth:
Ecofeminism in the Eighties'. This conference, which was
concerned with exploring the connections between feminism,
ecology and militarism, inspired the development of an ecofeminist network of women's groups and individuals in both the USA
and Great Britain.[2] Since then, a number of anthologies of ecofeminist writing have been published, and the movement has
developed several different, sometimes even contradictory,
emphases.[3] However, common to all types of ecofeminism is a
founding 'critique of male domination of both women and nature
and an attempt to frame an ethic free of male-gender bias'.[4] The
'woman–nature connection',[5] which ecofeminists have found to be
inherent in western culture, is a point of controversy. Some
ecofeminists embrace this concept as empowering and find it gives
women a particularly privileged stance in the rescue of the planet;
others reject the connection as a stifling patriarchal construct
which needs to be deconstructed. Ecofeminism found a popular
literary voice in the American author Marge Piercy, who, with
her 'eco-topian' novel *Woman on the Edge of Time*, inspired

readers to envision a radically different ecological society free of gender bias.[6]

Irmtraud Morgner, unlike Marge Piercy, did not live and work in a social context in which radically ecological theories were being discussed, and in all probability did not follow the Anglo-American ecofeminist debates. However, with *Amanda*, she penned a novel which ecofeminists may well want to keep on their bedside tables. This 'witch novel' (to follow the subtitle) is based on the central ecofeminist idea that the global threat is somehow linked with the relationship between the sexes, or rather, with the social construction of gender and the ensuing patterns of thought and social structures:

> Only if the other half of humanity, the women, bring into grand politics certain abilities and virtues, hitherto only developed for private purposes, can nuclear and ecological catastrophes be averted. Only if men, and the progressive governments led by men, first realize that they cannot cope with problems of world politics and ecology (and their own problems) without certain abilities and virtues of women, and, second, act accordingly, can the planet be saved. (*A*, 306)

Such are the authoritative claims of the siren Katharina, one of the characters in the novel. It is Catherine the Great who is reborn as a siren and called to the rescue of the earth, as is the protagonist Beatriz. With its recourse to Greek mythology, in this case the myth of Odysseus, Morgner's novel joins a group of GDR texts, mostly written in the seventies and eighties, which took up these myths with the intent of criticizing western civilization. New readings of the old myths were attempted in order to suggest social and cultural alternatives. Christa Wolf's novel *Cassandra* can serve as one example; it will be drawn upon for comparison below.[7] Heiner Müller, Franz Fühmann and Stefan Schütz are other GDR authors who have used Greek myths as a vantage point for wide-ranging critiques of modernity,[8] striving in the process for a modern 'thinking-myth-to-an-end'.[9] Any such reception of mythology, which for Marx cannot be a source of real knowledge, gains specific relevance in the GDR context.[10] By revisiting the roots of western civilization for their critical material, these authors imply an unbroken line of development of western civilization up to GDR society, and thereby negate the postulated essential difference between real-socialist society and capitalist western societies. It is this kind of approach which enables authors like Morgner to represent the ecological crisis as a result of an ideology underlying both capitalist and non-capitalist systems. As is clear in the above quotation, Morgner's sirens are needed to influence 'men' in 'grand politics' who have steered roughly the same (self-)destructive course in all political systems based on 'the myth of eternal economic growth' (*A*, 290).

Many different myths (northern European as well as Greek) have been worked into *Amanda*. However, the central myth, in which Morgner has her narrator Beatriz locate the utopian vision, is the myth of Pandora. In her reading of this myth the author encodes what I will argue is an ecofeminist standpoint. For according to Morgner, working with the Pandora material demands two fundamental decisions. On the one hand, it must be decided whether the potential inherent in the social construction of femininity is threatening or salutary: Pandora's box, she says, is a vessel, 'the belly of which is ultimately none other than the female belly; and, depending on the esteem in which woman is held, from this vessel flow all gifts or all ills'(*A*, 210). On the other hand, the myth demands a decision about 'the supposed victory of salvation over ruin – about the salvageability of the world' (ibid.).

I read Morgner's novel, with its decision in favour of the salutary potential of the culturally feminine, as a feminist 'green' text. In what follows, I intend first to present her reception of mythology as an example of the ecofeminist literary critique of western civilization, and then to pursue further key ecofeminist topics through the novel.

Amanda, the second part of Morgner's Salman trilogy, strikes the reader with its sheer length and complexity. Heterogeneous material – lectures, speeches, interviews and quotations – is interspersed throughout the book. Also there are numerous subsidiary plots. However, two main narrative levels can be identified: first the frame-narrative around the narrator of the book, the reborn Siren Beatriz, and secondly Beatriz's narration, the actual 'Book Amanda', in which she reconstructs the life of her former minstrel Laura Amanda. Wide-ranging responses to the global ecological and nuclear threat are given on both levels.

The fictitious narrator Beatriz was reborn as a siren, that is, as one of those mythological beings who threatened the safety of the

Homeric hero Odysseus with their song. Morgner's narrator, who is trying to grasp her own role, points to the other, derivative meaning of the word *siren* and wonders 'why a device used as an alarm signal would be given, of all names, the name "siren"'. She asks whether subconsciously 'the sirens are being remembered as alarming beings' after all, 'as warning criers' (*A*, 78).

Odysseus, who escaped the call of the Sirens, is considered to represent the prototypical European,[11] while the *Odyssey* is seen as an identification myth of western civilization. Adorno, for example, uses it as such in the *Dialectic of Enlightenment* when he describes the victory of instrumental reason over myth and the genesis of the patriarchal western societies. The sirens, 'with the irresistible promise of pleasure as which their song is heard, ... threaten the patriarchal order ... ', and they threaten work and progress.[12] The song of the Sirens is so powerful that it calls for violent countermeasures, like the elimination of hearing and the chaining up of the body. According to Adorno, 'measures like those taken on Odysseus' ship in regard to the Sirens form presentient allegory of the dialectic of enlightenment'.[13]

It is this archetype of occidental cultural history and this tradition of cultural criticism that Morgner evokes by personifying her narrator as a siren. And by redefining the sirens, no longer as a threat to humanity but, on the contrary, as beings who warn against danger, she calls into question the basic principles of our civilization. To purposeful Odysseus, the hero struggling for civilization through mastery of self and nature, what in Morgner's novel is defined as a necessary warning, a prophetic salutary message, seems bewitching and paralysing, as a diversion from his course.

What is the nature of the sirens' song? They sing of things past, for they know 'everything that ever happened on this so fruitful earth', and their allure is 'that of losing oneself in the past' which they are conjuring.[14] The compulsion 'to rescue what is gone as what is living' impedes the progress of enlightened Man, who must not use the past but as 'practicable knowledge', as 'the material of progress'.[15]

By contrast, salvaging the past is one of Morgner's principal concerns. To her, immersing oneself in the past offers the chance of finding oneself, not the danger of losing oneself. It is the ability to reflect and recall, to treat what is gone as what is living, that human beings, especially the (predominantly male) sustainers of civilization are lacking. She expresses this idea by tying the potential of the warning voice metaphorically to the capacity for recollection: because of the continuous, deafening noise of war throughout human history all sirens have lost their memory and therefore also their voices. This is why the narrator Beatriz needs to revive and train her memory – and for that purpose writes the 'Book Amanda'. She writes this book to remind herself, but also in order to remind her readers of something. Representative of the intended audience is Laura's son Wesselin, who, typically for his generation, 'seemed to master the facticity of the present to a surprising degree – at the price of a humanistic understanding of his origins' (A, 470). She hopes that reading the 'Book Amanda' 'can wean the socialist expert Wesselin a little of the conviction that only with his advent did the world embark on the course of efficiency and reason' (ibid.). Wesselin, who cannot 'muster any interest for things lacking screws', finds the idea of a 'land of milk and honey without switchboards and robots'(A, 429) inconceivable. His ideal landscape is Berlin Alexanderplatz. The fictitious addressee is thus clearly invested with Promethean traits. In fact it is not Odysseus but Prometheus to whom in Morgner's vision the sirens' voices are addressed. With this warning cry to Prometheus 'a long tradition of enthusiastic identification with the Titan in the history of Enlightenment thinking and Marxism' comes to a halt.[16]

The obscure oracle encoding the role of the resurrected sirens contains the words 'hope', 'Prometheus', 'Pandora', 'return' and 'song'. In order to decode the oracle, Beatriz undertakes to research the Pandora myth. She finds several interpretations which contradict the generally familiar (Hesiodian) version of seductive Pandora with her baneful box, and prudent Prometheus, who rejected the dangerous gift of the gods. First a female friend, a Greek fighter of the Resistance, tells her alternative creation myths with a strong feminist bias, in which Pandora appears as a gift from Gaia and the warning against the dangerous box is interpreted as a 'rumour' invented by 'revenge-plotting Zeus' (A, 66). Later she discovers Goethe's festival play The Return of Pandora.[17]

After years of working with the material of the Prometheus myth Goethe shifted the focus away from the Titan in this dramatic fragment. Prometheus, 'the busy man'(P, 29), represents the vita activa and is contrasted with his 'care-ridden, gravely pensive' brother Epimetheus (P, 314), representing the vita contemplativa. Talking about Pandora – for Prometheus 'the dangerous one' (*A*, 578) whom he rejected, for Epimetheus the 'heavenly one' (*P*, 579) whom he welcomed – they find themselves in a discussion about the essence of love, of happiness and about the concept of possession. To Prometheus – whose human race, the 'users' (*P*, 227), strives to exercise dominion over the earth and other peoples – only tangible goods are worth desiring: 'A man appropriates treasures daily with his fist' (*P*, 584). Grasping the world in this fashion, he proved unable to unite himself with the 'allgifted, allgiving' (this is how Goethe translates the Greek name Pandora) and indomitable woman. Epimetheus, however, understood how to find fulfilment through true devotion: 'Giving myself to her, that self I made my own' (*P*, 652). With great longing he preserves Pandora's memory and thus keeps alive the hope for her return.

Morgner's narrator is struck by what she understands as the play's central message and achievement: it turns longing, the 'hope backwards' (*A*, 210), into forward-looking hope and courage. She finds exemplary the new reading of the myth, in which the monstrous act of Pandora is 'reinterpreted as a cultural feat' (*A*, 78). Her return is, as it were, 'a return from prehistory' (ibid.), which could ring in a new phase of human evolution, if the predominant 'Promethean' world-view can be overcome. On the basis of her reading, research and reflection, Beatriz arrives at the following interpretation of the oracle, which captures metaphorically not only the causes of the present global situation and possible ways out of it, but also the role of the siren-narrator:

> Prometheus cannot leave under his own power the tracks of his thought, which have led him to admirable achievements and which now are steering him towards self-destruction. Sirens must put him off his course. Taken out of himself by their song, he will remember Pandora unobstructed by rumours. And he will be enabled, for the first time, to grasp himself and his work as a fragment and without future ... And the fourth human race, conceived with love by these two, could be the first to prove capable of peace ... (*A*, 129)

Here the plea for recollection, which connects the new reading of the siren myth with the Pandora version as derived from Goethe, reveals its chief implications. To 'remember Pandora unobstructed by rumours' means going back to versions preceding the familiar ones based on Hesiod, which are here dismissed as a misogynist 'rumour'.[18] The point of this digging into prehistory is therefore to strip the myths of their later patriarchal interpretations, to trace them back to their 'matristic substrata'.[19] The point is not that these layers are 'more authentic', but that the process of uncovering shows the myths (through which a civilization interprets itself) in their historical specificity.

Comparable reconstructions of pre-patriarchal mythical strata were attempted by Christa Wolf in *Cassandra* and Stefan Schütz in *Medusa*. The purpose in both instances is similar to Morgner's: by having recourse to prehistory, they are on the one hand critiquing the European type of civilization and also as it was continued in the GDR; on the other hand, they are locating the beginning of a (self-) destructive development in the transition from matriarchal to patriarchal cultures. Mythical female figures like Medusa, Medea, Cassandra, Pandora or the sirens, who in the familiar versions represent danger and ruin, are rehabilitated and interpreted as originally beneficial voices suppressed and defamed in subsequent cultural history. The beginning of the misdirected development, it is suggested, coincided with the defamation of the feminine.

Promethean thought, as it is characterized by Goethe and Morgner, knows no alternative to either domination or submission, which means that the gender dichotomy, too, can only be conceived in hierarchical terms. Goethe's Prometheus created woman 'with forethought of the man whose servant she would be' (*P*, 593). The marginalization of the feminine, which marks the beginning of the patriarchal order, is written into the myths. These authors are suggesting that it is now time to 'learn to read myth'[20] in order to expose the legitimating strategies of dominant ideologies. This is not meant as a plea for matriarchy, but rather for change. The oldest, matriarchal layers of myths reveal alternatives, which remind us that civilization has not always been this way, and that it therefore does not have to remain as it is. The view back to prehistory is supposed to open up a window on possible futures: myth yields a glimpse of utopia.

For Morgner's narrator this is the utopian vision of Pandora's return, eagerly awaited by Prometheus once he is prepared to unite himself with her in love and equality. This means: one day Promethean thinking will open up towards the unmeasurable, unpossessible, and will be able to embrace the hitherto marginalized other. Only the end of hierarchical dualism, which has served to legitimate the dismissal of any 'other' – be it woman, nature or non-European peoples – will succeed in calling a halt to the threatening (self-)destruction of humans and their environment – this is what Morgner arrives at by 'thinking the myth to an end'.[21]

The connection between androcentrism and environmental destruction, as well as the call to abandon the principle of domination, are fundamental tenets of ecofeminist thinking.[22] Especially the more spiritualist strands of ecofeminism also like to borrow from mythology; the personified 'mother Gaia' figures in several discourses.[23] However, socialist feminists, for example Janet Biehl, warn against employing myths in alternative political movements: 'Myth cannot fight myth. Ruling classes have always encouraged confusions between illusion and reality . . . In an age of manipulation and myth . . . we must restore to the ecology movement a realistic – not illusory – view of nature, and a political – not a religious – view of politics.'[24]

Neither Morgner nor the other GDR authors use myths to cast a veil of illusion over reality, but rather to deconstruct myths and the mythical canon. According to Wolfgang Emmerich, this literature 'does not found a new mythology, but reconstructs itself by descrying the old one'; that is, it attempts a 'reappropriation of what has been repressed to our detriment'.[25]

In avoiding the foundation of a new mythology, this literature points to a very sensitive (eco)feminist topic: the positive identification of women with 'mother Earth'. Such identification is advocated in the more spiritualist strands of the ecology movement, such as deep ecology, and leads to emotional campaigns replete with 'Love your Mother' T-shirts etc.[26] Approaches like this are being criticized, because they support the same essentialism they ought to be fighting:

> The idea that one group of people is closer to nature than others assumes the very nature–culture split that ecofeminism would deny. . . If we humans are essentially or naturally dichotomized by sex-linked traits, then there is a certain futility in trying to change human cultural practices.[27]

Morgner was well aware of this conceptual trap. In the chapter 'The apocalypse of Konrad Tenner' she sarcastically disowns maternalism. Predicting the nuclear and ecological apocalypse, Tenner, the male protagonist, pathetically abjures science and scholarship, only to end with the apotheosis of woman:

> For how does a man measure up to one who can give birth to human beings? ... Men had to condemn Nature as a whore and pregnant women as sullied in order to transform feelings of impotence into superiority ... Root impulse behind all male scholarship: a substitute for child-bearing (A, 219f.).

Tenner's wife Vilma recognizes the reactionary and covertly misogynist tendency of this woman-worship shining through the seemingly radical (and feminist) cultural critique and counters: 'And because you and your like, after the pleasure of thinking, now believe you have reached rock bottom, I am not supposed to even start? Should this pleasure be denied to my kind, only because the other kind believe they have thoroughly bungled it?' (A, 220).

Though instrumental reason and 'male' science may have pushed our civilization to the brink of catastrophe, Morgner does not see fit to establish a feminine counter-world of emotions and sensuality. She makes a point of portraying Laura and Vilma as rationally motivated, as readers and speakers. They are resourceful, witty women, more gruff than sweet. Laura is a dedicated train-driver. The author is careful not to lapse into culturally conservative figures of argument, and therefore does not pit male reason against female emotions, outwardness against inwardness, technology against nature or machine against organism.

This stance of Morgner's also finds expression in her narrative style. First, she chooses a detached perspective: Laura is shown in a very distant way and does not lend herself as easily as do most female protagonists created by women authors to identification. Furthermore, Morgner keeps to a historically specific reality, a fact which has been overlooked in secondary literature in favour of an emphasis on her fantastical elements. A comparison with *Cassandra* will prove illuminating here. Both novels try to develop a female-identified image to counter a way of dealing with the world which they define as patriarchal and destructive. In *Cassandra*, this counter-image is located in the community of Mount Ida. There, in caves by the river, live women and a few men of different ages, different social background and from both hostile camps, together in equality and solidarity. They make baskets, form pots from clay and spend the evenings squatting round the fire, telling tales. It is here that Cassandra finds healing, a new spiritual home and a counter-world to the war machine of Trojan society which is becoming ever more threatening. Wolf created a very evocative utopia of an unalienated life, close to nature and free of domination. However, it could equally be seen as a kind of 'eco-idyll', easily marketable as an alternative holiday for all those Europeans who are weary of civilization and technology, who feel alienated and strive for self-realization.

How, then, can people deal with their environment in a peaceful and constructive way, if they do not live in caves on the Scamander but rather on the thirteenth floor of a tower block in the city centre of East Berlin? This is the question with which Morgner grapples.

Remaining consistent with her analysis of the causes of the crisis, Morgner envisions the solution not as a movement 'back to nature' or as a set of concrete measures for protecting the environment and securing peace. Rather, she pleads for developing a new relationship with the environment. In an inserted speech, declared a lecture from the 'Blocksberg University', this relationship is described as 'atheist religion' (*A*, 373). This term is meant to signify a posited metaphorical and relational way of thinking, which according to the speaker had been suppressed in favour of abstract thinking for ages. The sheer extent of the catastrophes and devastations, she suggests, can only fit comfortably 'into a person's head as abstractions' (*A*, 374). A true understanding of our situation could therefore only be achieved by somebody, 'who does not understand the world solely as material, but who is also capable of feeling a part of it – bound, . . . sheltered, responsible' (ibid.). She continues to say that the 'metaphorical' way of appropriating the world can still be viewed in its results – in myths and religions – though the capability of appropriating the world to oneself in this fashion had been lost in most industrial countries and is only cherished by poets, if at all (*A*, 375). When the speaker concludes by pleading for the practice of personification, she is urged by the audience to offer 'crutch-philosophy' or 'ersatz religion' (*A*, 450). Nevertheless, Morgner grants her the last word: 'Even though it does not exist in nature, personification . . . can counteract the growing alienation of people in their mechanized environment. It can make us sense more clearly what it is we have to preserve from nuclear war and ecological destruction' (*A*, 451). It is obvious that the novel is here also interpreting itself and allocating to itself a role in the required process of recovery. With its mythical and fantastical elements it achieves a pronouncedly 'metaphorical appropriation of the world' in order to preserve this way of relating to the world, perhaps also to accustom the readers to it.

Of the characters in the 'Book Amanda' it is above all Laura's father, the train-driver Johann Salman, who communicates with his environment through the technique of personification:

> Salman's harmonious character was apparently based on his ability to establish human relationships not only with people . . . [I] remembered that he could not only talk to his machine, but also with the Chemnitz valley and the textile factory VEB Doppelmoppel and with the town Dietensdorf. From his driver's cab he would call out to Dietensdorf: Hey, old girl, looking pretty chilled through again this morning. (A, 493)

Dismissing gender essentialism, Morgner chooses a man to illustrate a way of 'knowing' the world which has been ousted by patriarchal scientific civilization. Similarly, Laura's son Wesselin, on the one hand the representative of Promethean socialist man, has, on the other hand, imaginative ways of 'making himself at home' in his environment. Laura feels alienated in her new tower block and laments the lack of 'nature' and 'history' until Wesselin takes her with him to explore the surroundings, digging up shells, bones and other finds, from which he constructs a history of the place (chapter 44). Laura often learns in this way from her son, who does not yet use categories to divide his world into nature and non-nature. She is baffled when Wesselin asks her to 'talk to Nature' (A, 428) on his behalf, as if nature were a realm only accessible to older generations. This raises the question of just how relevant Laura's traditional concept of nature as birds and trees is to a child of the big city; it is another example of Morgner's attempt to avoid essentialism and the nature–culture dualism.

The narrator calls Johann Salman's personal relationship with the environment 'eroticism towards the world' (A, 494). Something like this 'personifying eroticism' is necessary, for 'a human race that is incapable of loving its earth is likely to be also incapable of saving it from annihilation' (ibid.). Through her examples, Morgner emphasizes that 'the world' that people need to relate caringly to comprises locomotives and inner cities. Her critical concept of 'nature' and 'environment' echoes other critiques of over-simplified green or feminist thought.[28]

Eroticism against alienation has been proposed by several ecofeminists as a programme for the required process of personal and social change. It points to a central problem: the pessimistic view of human nature which dominates most eco-discourses. Normally the implicit and explicit appeals summon people to self-discipline and self-restriction – there are few people by now who do not respond to the topic of the ecological crisis with some sort of bad conscience. Yet is it not a perverted understanding of love of nature which conceives of this 'as a repression of a destructive desire, rather than as a release of human desire to participate creatively in the natural world'?[29]

Aggression, competitiveness and the striving for autonomy characterize an attitude which is privileged in our societies and is generally felt to be 'male' – it is, however, 'culturally cultivated, not naturally male' (*A*, 306), says Morgner. Traditional women's roles cultivate very different abilities. According to feminist social scientists our gender-specific social roles not only influence our social behaviour, but also our potential ways of knowing. Joan Tronto speaks of a 'tradition of relatedness': 'many women develop a relational way of loving and knowing informed by their direct experiences in caring for people of different ages, needs, and abilities.'[30] For Morgner this 'ability to care', which 'for several millennia the dominant specialists' culture has highly developed only in women' (*A*, 306), represents a potential for salvation. Privileging this ability appears a possible way out of the crisis: 'Suddenly we have arrived at the historical point when this ability becomes indispensable, on penalty of ruin, for the greatest public purposes' (*A*, 306, cf. also 445).

Herewith Morgner preaches what Judith Plant calls the central message of ecofeminism: '. . . that we all must cultivate the human characteristics of gentleness and caring . . . ', that, in a conscious act of choice, we must 'claim those aspects of our socialization that are of benefit to the species'.[31] Although not personifying 'gentleness' as it is usually understood (lest they remind the reader of the traditional ideal of meek femininity), Morgner's women characters do prove a lot more caring than their male counterparts, if caring is to be understood in terms of understanding, responsibility and solidarity. Throughout her book the author makes a point of showing how these characteristics and abilities are a product of the demands and problems her women characters face in their everyday lives. As one female train-driver colleague of Laura's explains it: her commendable responsibility and discipline cannot be put down to her being a 'character-athlete' but rather to her being a mother of four (*A*, 127). Laura's double role as train-driver on night shifts and mother of the boy Wesselin is shown to school her daily in altruism, the ability to compromise, responsibility and bonding. At another level, however, her situation as a working single mother is not presented as a positive ideal, but as a stressful overload:

> She mastered the second shift, in which she attended to the tasks of housewife and mother, only under severest self-discipline. She could spare no energy for playing with Wesselin. He reacted to this privation by pulling his hair out. His misery consumed the last bit of her will to live. (*A*, 127)

The problem of the second shift and other related problems, like women's compromising their careers because of motherhood, are traditional feminist concerns which Morgner revisits in several places.[32] Laura's case, however, encompasses a more far-reaching critique: despite fledgeling attempts to socialize reproduction – Wesselin's all-day childcare is an example – there remain still more housework and family duties than a working woman (or man) can master without overextending herself. The example of Laura and Wesselin demonstrates just how much time, energy and emotional commitment are involved in the upbringing of a child. Laura is his partner for conversation and play, his source of knowledge, love and security: she has to live up to demands which an institution can hardly satisfy.

In short, Morgner joins socialist feminists in exposing the non-public gender hierarchy: she renders visible the neglected sphere of reproduction. 'Women's responsibility for reproduction includes both the biological reproduction of the species ... and the intra-generational reproduction of the work force through unpaid labor in the home. Here too is included the reproduction of social relations – socialization.'[33] Shortly before the *Wende* [collapse of the GDR] the East German critic Irene Dölling pointed out 'the necessity of a theory of reproduction'. Without such a theory, she recognized, it would remain inexplicable why 'despite extensive ... measures taken for the improvement of the economic and social situation of women, traditional stereotypical patterns of a gender hierarchy, of a devaluation of all that is "feminine", do not disappear'.[34]

All this considered, Morgner's social analysis of the 'specialist culture' becomes clear: women traditionally carry the main burden of the reproduction of society – and the GDR's actually existing socialism had failed to change that. As long as a whole category of work and responsibility and its corresponding abilities and values – e.g. 'the ability to care' – are banished to the private sphere, and thus remain invisible, unsupported and ineffectual in the public realm, the patriarchal gender hierarchy will not be abolished. And neither will a challenge be raised to the hierarchically dualist relationship between production and reproduction, a relationship which destroys the environment.[35]

Morgner's siren offered the prophecy (quoted above) that ecological disasters can be prevented 'only if the other half of humankind, the women, bring into grand politics certain abilities and virtues hitherto only developed for private purposes' (*A*, 306). Translated into the terminology that befits Morgner as a Marxist thinker, this can be read as a demand for the recognition of the reproductive sphere. It is this demand that most fundamentally connects feminist and ecological concerns. Both nature and women have served as cheap (often over-exploited) resources allowing progress and economic growth for patriarchal industrial societies. In both cases the unrewarded sacrifice has been left out of calculations of profit and progress. Western socialist ecofeminists, too, have come to the conclusion that we need to make 'the category of reproduction, rather than production, central to the concept of a just, sustainable world'.[36] Thus the much-heralded 'ethic of caring' acquires an underpinning in social theory.

In using myth critically to 'demystify' patriarchal culture; in refusing to link woman with 'nature' in her imagery and choice of settings; in avoiding essentialism in her discussion of gender and ecology; and in rooting her suggestions for alternative ways of living firmly in late twentieth-century reality rather than unrealistic utopian settings, Morgner demonstrates a subtle and sophisticated grasp of the issues that have occupied western ecofeminists. Demonstrating an affinity with such a decidedly western apocalyptic movement, Morgner assumes a 'symmetry of self-destructive tendencies in east and west' and a 'negative convergence of the systems',[37] an assumption that according to Emmerich brought many GDR texts closer to West German readers' experience.

Within the specific context of ecofeminist thought, Morgner in fact precedes the development of the topic in West German literature. Given her socialist background, it should not be surprising that Morgner is able to relate environmental and gender exploitation. Socialist analysis has always aimed to give voice to previously invisible historical forces. With Marx it was the class of working men fuelling industrial production. As the social and environmental costs of industrialization have grown ever more difficult to ignore in this century, it was practically inevitable that a way of thinking attuned to unrewarded exploitation would notice also the work of women and the plundering of the environment. From a socialist view of production, in other words, it is a short step to the sphere of reproduction. For Morgner and for socialist ecofeminists a revaluation and retheorization of reproduction must go hand in hand with a transcendence of the west's most cherished hierarchical dualisms. The reign of progress, productivity and property (personified by Prometheus and Odysseus) over stasis, preservation and memory (represented by Epimetheus and the sirens), of male over female, of public over private, of war over peace, must yield to an egalitarian relational thinking, free of domination. These dualisms, once again, are fundamental to industrial socialist as well as capitalist societies. Thus Morgner's vision, like that of other recent GDR authors rereading western myths, remains relevant after the demise of east European communist governments.

Notes

[1] Irmtraud Morgner, *Amanda. Ein Hexenroman* [*Amanda: A Witch Novel*] (Hamburg and Zurich, 1984), 290. Subsequent references will be indicated by the abbreviation *A*. This and all further quotations either from *Amanda* or from other texts unavailable in English appear in my own translation.

[2] The conference took place in Amherst, Mass., in March 1980. See Leonie Caldecott and Stephanie Leland (eds.), *Reclaim the Earth: Women Speak out for Life on Earth* (London, 1983), 4–8.

[3] See, for example, Caldecott and Leland (eds.), *Reclaim the Earth*; Judith Plant (ed.), *Healing the Wounds: The Promise of Ecofeminism* (Philadelphia, 1989); Irene Diamond and Gloria Orenstein (eds.), *Reweaving the World: The Emergence of Ecofeminism* (San Francisco, 1990); Greta Gaard (ed.), *Ecofeminism: Women, Animals, Nature* (Philadelphia, 1993); Karen Warren (ed.), *Ecological Feminism* (London

and New York, 1994).

4. Philosopher Karen Warren, quoted in Carolyn Merchant, *Radical Ecologies: The Search for a Livable World* (London and New York, 1992), 185.

5. See, for example, Ynestra King, 'Healing the wounds: feminism, ecology, and the nature–culture dualism', in Diamond and Orenstein (eds.), *Reweaving the World*, 117. This issue is addressed in almost every ecofeminist essay.

6. Marge Piercy, *Woman on the Edge of Time* (New York, 1977). Poetry by Piercy was included in the anthology *Reclaim the Earth*, ed. Caldecott and Leland, 89.

7. Christa Wolf, *Cassandra: A Novel and Four Essays* (New York, 1984).

8. See, for example, Franz Fühmann, *Der Geliebte der Morgenröte* (Rostock, 1976) and *Das Ohr des Dionysos* (Rostock, 1985); Heiner Müller, *Verkommenes Ufer, Medeamaterial, Landschaft mit Argonauten* (Bruchsal, 1987); Stefan Schütz, *Antiope und Theseus (Die Amazonen)* (1974) published in Schütz, *Heloisa und Abaelard* (Berlin, 1979), and *Medusa* (Reinbek bei Hamburg, 1986).

9. Wolfgang Emmerich, 'Zu-Ende-Denken. Griechische Mythologie und neuere DDR-Literatur', in *Kontroversen, alte und neue. Akten des VII. Internationalen Germanistenkongresses, Göttingen 1985*, 10 (Tübingen, 1986), 217. Emmerich coined this term ('das Zu-Ende-Denken' des Mythos) after Hans Blumenberg's 'Den Mythos zu Ende bringen'. In his paper he discusses Müller's play *Philoktet* as an early, prototypical example.

10. Cf. Karl Marx, *Critique of Political Economy*, ed. Maurice Dobb (New York, 1970), 216.

11. Cf. for example Heiner Müller, 'Letter to the director of the Bulgarian première of *Philoktet*', in *Herzstück (Texte 7)* (Berlin, 1983), 108. I was alerted to this text by Emmerich, 'Zu-Ende-Denken', 222.

12. Max Horkheimer and Theodor W. Adorno, *Dialectic of Enlightenment*, tr. John Cumming (New York, 1987), 33.

13. Ibid., 34.

14. Adorno's translation of Homer, as translated by Cumming, *Dialectic of Enlightenment*, 33 and 32.

15. Ibid.

16. Wolfgang Emmerich, *Kleine Literaturgeschichte der DDR* (Frankfurt am Main, 1989), 281 (my translation). Emmerich made this claim originally about Braun's poem 'Verfahren Prometheus' (1982), but his quotation evidently describes a more general trend in GDR literature.

17. Morgner uses the working title of Goethe's play, which was published in 1810 as *Pandora: A Festival Play*. All quotations are indicated by the abbreviation *P*, and are taken from Johann Wolfgang Goethe, *Verse Plays*

and Epic, ed. Cyrus Hamlin and Frank Ryder, tr. Michael Hamburger, Hunter Hannum and David Luke (*Goethe's Collected Works*, 8) (New York, 1987).

18. Compare with the findings of historical mythologists such as Jane E. Harrison. Pandora was originally the Earth goddess, who was then 'changed and minished' in the 'patriarchal mythology of Hesiod'. *Prolegomena to the Study of Greek Religion* (Cambridge, 1903), 284.

19. Morgner in an interview with E. Kaufmann, in Marlies Gerhardt (ed.), *Irmtraud Morgner. Texte, Daten, Bilder* (Frankfurt am Main, 1990), 47 (my translation).

20. Christa Wolf's postulate. Wolf, *Cassandra*, 196.

21. Compare with western feminist critics of science, e.g. Carolyn Merchant, *The Death of Nature: Women, Ecology and the Scientific Revolution* (New York, 1980); Evelyn Fox Keller, *Reflections on Gender and Science* (New Haven, 1984).

22. One of many examples is Janis Birkeland, 'Ecofeminism: linking theory and practice', in Gaard (ed.), *Ecofeminism*, 13–59. She sees ecofeminism as offering 'a political analysis that explores the links between androcentrism and environmental destruction' (p.18).

23. Cf., for example, Riane Eisler, 'The Gaia tradition and the partnership future: an ecofeminist manifesto', in Diamond and Orenstein (eds.), *Reweaving the World*, 23–34.

24. Janet Biehl, 'The politics of myth', *Green Perspectives* (June 1988), 28.

25. Emmerich, 'Zu-Ende-Denken', 223 (my translation).

26. Cf., for example, Chaia Heller, 'For the love of nature: ecology and the cult of the Romantic', in Gaard (ed.), *Ecofeminism*, 224.

27. Joni Seager, *Earth Follies: Feminism, Politics and the Environment* (London, 1993), 246f.

28. For environmental philosophy see Ingolfur Blühdorn in ch.5 of this volume; for feminist theory see, e.g., Donna Haraway, *Simians, Cyborgs, and Women: The Reinvention of Nature* (New York, 1991).

29. Heller, 'For the love of nature', 227.

30. Ibid., 233, summarizing Tronto.

31. Plant, *Healing the Wounds*, 3.

32. See, for example, pp. 155, 184, 279f., 283 etc.

33. Sociologist Abby Peterson, quoted in Merchant, *Radical Ecology*, 197.

34. Irene Dölling, 'Marxismus und Frauenfrage in der DDR. Bemerkungen zu einer notwendigen Debatte', *Das Argument*, 177 (1989), 716.

35. Cf. *A*, 290.

36. Merchant, *Radical Ecology*, 195.

37. Emmerich, 'Zu-Ende-Denken', 223 (my translation).

Part III
Post-War German Literature

9 • Catastrophism in Post-war German Literature

Axel Goodbody

Perhaps the most striking characteristic of the cultural climate of the Federal Republic in the early 1980s was the extraordinary pessimism and the fascination with natural and man-made catastrophes pervading both popular and high culture. Under the heading 'Katastrophismus' Hermann Glaser wrote at the end of the decade in his *Kulturgeschichte der Bundesrepublik Deutschland 1968–89* [*Cultural History of the Federal Republic of Germany 1968–89*] of the prevailing atmosphere among West German intellectuals and in contemporary youth culture as:

> A certain masochistic tendency to self-negation ... being prepared to give up the sovereignty and autonomy of the individual in favour of 'objective' forces and powers scarcely capable of definition, interpretation or control ... Stylization of one's own wretched situation and cultivation of historical pessimism have become the dominant characteristics of our culture ... 'Catastrophism' has long since become a substitute for ideology and a pseudo-religion with false progressive claims.[1]

Literature and non-fiction on man's degradation and destruction of the natural environment, the dangers facing mankind through over-population, the extinction of species, pollution, and the horrific consequences of a possible nuclear war painted an overwhelmingly negative picture of the history of mankind, culminating more often than not in a disaster wiping out modern civilization.

An early dissenting voice was the left-wing essayist Michael Schneider, brother of the novelist Peter Schneider, who reacted allergically to the bleakness of contemporary novels and poems, films and popular philosophy. His stimulating articles on 'Apocalypse, politics as psychosis and the playboys of doom' and 'The intellectuals and catastrophism: crisis or turning-point among German Enlightenment writers?'[2] do not deny the objective grounds for a sombre assessment of the political and ecological situation, but take issue with the fatalist, sometimes even nihilist frame of mind of his fellow writers. Seeking explanations for the phenomenon, Schneider suggests that catastrophism is an expression of anxieties stirred up by a time of fundamental change, and of the breakdown of those values which normally help allay individual fears, i.e. the promise of peace and employment, economic growth, social security, progress and technology. Finally, he gives a critical assessment of the merits of some individual works as contributions to public debate on the danger of nuclear war and ecological deterioration.

In the second half of the decade two extended academic studies of catastrophism took up Schneider's theme, situating literary expressions of the contemporary West German *Zeitgeist* in the context of international apocalyptic writing and film,[3] and examining its roots in the biblical apocalypses and nineteenth- and early twentieth-century German thought.[4] Jost Hermand's survey of 'Green utopias in Germany',[5] which reviews German cultural tradition from the standpoint of utopian and dystopian visions with an ecological slant, juxtaposing literature with political and philosophical statements, contains further insights, but is disappointingly brief in commenting on individual literary works. More detailed information on catastrophist prose and poetry in the seventies and eighties may be gained from Reinhold Grimm's pioneering article 'The ice age cometh',[6] Paul Konrad Kurz's collected reviews in the volume *Apokalyptische Zeit* [*Apocalyptic Times*],[7] Volker Lilienthal's article 'Will o' the wisps from the darkness of the future'[8] on recent German catastrophe literature, as well as from articles on individual writers.[9] Volker Lilienthal exemplifies a more detached approach to the catastrophist literature of the eighties than Glaser or Schneider. He distinguishes between different strands of writing reflecting shallow sensationalism, loss of hope in a future worth living in, and the operative response of a literature concerned with contemporary issues to the challenge presented by events such as the NATO decision to station missiles with atomic warheads on German soil in 1979, or the Chernobyl nuclear disaster in 1986. On the whole, Lilienthal sees writing in the catastrophist genre as attempting to shake readers out of their indifference. In an appendix he lists over a hundred contemporary German novels, poetry volumes, plays, radio plays and films on the theme of world catastrophe. Besides the unashamedly playful, entertaining approach of such volumes as Ingomar von Kieseritzky's *Buch der Desaster* [*Book of Disasters*] or Georg Lentz's novel *Der Herzstecher* [*The Heartsting*], and the more thought-provoking but still relatively slight novels of Matthias Horx (*Es geht voran* [*Making Progress*] and *Glückliche Reise* [*Happy Journey*]), Anton-Andreas Guha (*ENDE. Tagebuch aus dem 3. Weltkrieg* [*END. Diary from the Third World War*]) and Rüdiger Hipp (*Grandhotel Abgrund* [*Hotel The Abyss*]), and Gudrun Pausewang's prize-winning novels for teenage readers (*Die letzten Kinder von Schewenborn* [*The Last Children of Schewenborn*] and *Die Wolke* [*The Cloud*]), Lilienthal discusses aesthetically more ambitious works including Christa Wolf's *Kassandra* [*Cassandra*] and *Störfall* [*Accident*], Günter Grass's *Die Rättin* [*The Rat*] and the Austrian writer Inge Merkel's *Die letzte Posaune* [*The Last Trump*].

A comprehensive overview of the dystopian catastrophist strand in post-war German literature is beyond the scope of this chapter. After sketching out the political and ideological context of catastrophism in the eighties, and its roots in German tradition, I shall therefore restrict my comments to some of the more significant works in the genre, focusing on the following two questions: What role has catastrophism played in promoting ecological awareness? And how successfully has the ecological message been reconciled with literary quality?

The ideological background to catastrophism in the mid-eighties may be illustrated by reference to two philosophical best-sellers arguing man's collective self-destruction was unavoidable. These were the Münster philosophy professor Ulrich Horstmann's *Das Untier. Konturen einer Philosophie der Menschenflucht* [*The Monster: The Parameters of a Philosophy of Flight from Mankind*] (1983)[10] and Hoimar von Ditfurth's *So laßt uns denn ein Apfelbäumchen pflanzen. Es ist soweit* [*Then let us Plant an Apple Tree: It is Time*] (1985).[11] Horstmann reviews the philosophy of Voltaire, d'Holbach, Klages, Freud, Foucault, Günther Anders and E. M. Cioran, arguing, apparently in all seriousness, for a nuclear war as soon as possible, to end human civilization. For mankind is an evolutionary error, programmed to self-destruct. Only by means of nuclear war can peace and order be restored to the universe. The central argument in von Ditfurth's less abstract study, which presents an impressive barrage of environmental facts and figures, is also that mankind will soon be extinct, either as a result of nuclear self-annihilation or through destruction of the biosphere (the latter mainly a consequence of over-population). Von Ditfurth also believes mankind is genetically predetermined, but differs from Horstmann in concluding we should learn to accept the probable end of our species at some time in the future as a natural event, and one which does not render human life meaningless.

The external factors which gave rise to a veritable spring tide of apocalyptic texts around 1983 are readily identified as the threat of nuclear war and public 'discovery' of acid rain and forest die-back. Brinkmanship between the superpowers in the late seventies had led to the NATO twin-track decision to station American medium-range nuclear missiles in western Europe, and the intense public debate, which gave birth to the peace movement and was instrumental in the electoral breakthrough of the Green Party, came to a head in the 'hot autumn' of 1983, when a majority of members of the Bundestag voted in favour of allowing Pershing II missiles to be stationed in Germany. At the same time press reports about forest die-back and the unprecedented deterioration of German forests began to appear. The term *Waldsterben*, coined by *Der Spiegel*, whipped up powerful emotions. Industrial and traffic pollution were soon identified as the most important factors in a complex of causes. Other man-made catastrophes contributing to a deeply pessimistic view of the future at the time included accidents in chemical factories (Seveso and Bhopal), the accident in the US nuclear power station Three Mile Island (Harrisburg, Pennsylvania), alarming reports on the rapidly increasing disappearance of species, on the hole in the ozone layer, the greenhouse effect, the population explosion and famine in the Third World. However, these concerns, which may be classed as ecological in the broader sense, do not on their own explain the propensity to catastrophism in German writing. They were accompanied by more general underlying political and social factors: disillusionment with socialist ideals on the left after the petering out of the student movement and the expatriation of Wolf Biermann from the GDR, and anxieties over signs of the emergence of a police state in 1977 in response to the terrorist threat. Catastrophism thus reflects a social and cultural crisis rather than a merely ecological one.

To English readers, catastrophism may appear a particularly German preoccupation. Klaus Vondung notes that fears of the end of the world and awareness of the dangers facing mankind have been especially strong in modern Germany, and contrast with wide-spread optimism about the future in the USA, as well as the enthusiasm of most French intellectuals for technology, seen in their relative indifference to the Chernobyl accident.[12] Debate in Germany is unique in its intensity and the use of apocalyptic imagery. Jost Hermand also remarks on the dearth of post-war liter-ary utopias in Germany, and the comparative wealth of dystopias. Michael Schneider suggests some plausible reasons why German intellectuals were particularly affected by the international phenom-enon of a crisis of Enlightenment optimism and faith in progress in the seventies.[13] Melancholia resulted in Germany when a generation who had spent their childhood and youth in the stifling and joyless climate of sterile post-war reconstruction, and resorted in conse-quence to dogmatic abstraction, found their beliefs undermined. Schneider also suggests that the depressive disposition of the post-war generation may be explained as an involuntary reaction against the heroic reconstruction stance of their fathers and mothers, who had suppressed melancholy and resignation in self-protection. Hence Germany is characterized by a nihilism based on weariness with the world in the midst of affluence, reflecting the disturbed mentality of a nation of aggressors and losers.

Whatever truth there may be in this, catastrophism in the seven-ties and eighties also has undoubted links with the German tradition of cultural pessimism exemplified by Friedrich Nietzsche, Ludwig Klages and Oswald Spengler. And apocalyptic literature in Germany reaches back beyond Expressionist drama to Wagner's *Götterdämmerung* [*Twilight of the Gods*], and beyond the poetry of the First World War to that of the Napoleonic Wars of Liberation. These make extensive use of archetypes of destruction such as the Flood, the Tower of Babel and Sodom and Gomorrah in the Book of Genesis, and the legendary fall of Troy. Much of the fascination of the various Old and New Testament apocalypses derives from their poetic qualities, and the biblical images of fire, floods, earth-quakes, the sun darkening, freezing cold, hail, a rain of ash, etc., are to be found in works as different as Expressionist poems and novels on nuclear winter written in the eighties.

The traditional function of apocalyptic writing has been to help give meaning to a period of suffering by suggesting that it will come to an end. The Revelation of St John, which predicted the overthrow of Roman persecutors, the resurrection and judgement of the dead, and the establishment of a Heavenly Jerusalem populated by all true believers, were written to strengthen the hope and determination of the infant church. The apocalyptic world-view generalizes specific historical oppression by giving it moral, cosmic and mythical dimensions. On a more problematic level, the emphasis on the chaotic, evil, moribund state of the present serves to justify the unleashing of destructive and aggressive energy in compensatory wish-fulfilment. As Klaus Vondung reveals, apocalyptic thinking possesses political and social as well as religious, aesthetic and existential dimensions, and the last two centuries have witnessed a particularly ill-fated link between apocalyptic thinking and German nationalism.

In most modern apocalyptic writing the disaster is no longer held to be the work of a divine agent, but rather that of man. Vondung makes a further distinction between traditional millenarian apocalypses and the modern 'kupierte Apokalypse' [truncated apocalypse].[14] Whereas in the Book of Revelation the end of the world was only a transitional phase, en route to the New Jerusalem, such visions of renewal and rebirth have become rare since the late nineteenth century. The apocalypse continues to satisfy a desire to give symbolic shape to the problems and events preoccupying us, and to give them meaning, but now in the form of a warning of the terminal consequences of our present behaviour. Ecologically motivated catastrophism thus has roots in apocalyptic tradition, but differs significantly from it in denying its millenarian vision.

There have been three main phases of catastrophism in twentieth-century German culture. These can be located chronologically around the First World War, after the Second World War, and from the mid-seventies to the mid-eighties. In the years leading up to the First World War, Georg Heym, Jakob van Hoddis and Georg Trakl revealed in their poems a fascination with death, decay and doom. 'Weltende' [the end of the world], the rallying-cry of the Expressionist generation, served, broadly speaking, as a metaphor for the end of bourgeois society. In the poem 'Der Krieg' ['War'], Georg Heym presented war as a cataclysm fit to purge Germany of the stifling restrictions and injustices of Wilhelmine society. Virtually indistinguishable from revolution, it was felt as a liberating force, and associated with storm and natural catastrophe. A spate of optimistic apocalyptic works formulating visions of a new world and a new mankind followed at the end of the First World War: Ernst Toller's play *Die Wandlung* [*The Transformation*] (1919) and the poems in Kurt Pinthus's anthology *Menschheitsdämmerung* [*Twilight of Mankind*] (1920) are among the best-known.

The trauma of the collapse of the Third Reich and the revelation of the Holocaust, and the situation of hunger, unemployment and homelessness immediately after the Second World War triggered off a new flood of apocalyptic texts which have been interpreted in psychoanalytic terms as collective displaced fantasies of guilt and punishment. Hermann Kasack's novel *Die Stadt hinter dem Strom* [*The City Beyond the River*] (1946) is typical in that the catastrophe is portrayed as an act of fate rather than of human doing. Human existence is denied inherent meaning. Though the catastrophist world-view exercised less fascination over Germans as post-war reconstruction progressed and the economy improved, apocalyptic writing continued from the early fifties into the mid-sixties, now often reflecting anxieties over the atom bomb and fear of a third world war as well as the experience of the past war. Historical pessimism and fear of a new catastrophe, expressed in 1951 in Arno Schmidt's stories *Brand's Haide* [*Brand's Haide*] and *Schwarze Spiegel* [*Black Mirrors*],[15] reach a peak in Schmidt's work in *Kaff auch Mare Crisium* [*Kaff Also Mare Crisium*] (1960),[16] which harks back to the world conflagration (Ragnarök) in Norse mythology, and alludes to the Burgundians' death by fire in the *Nibelungenlied* [*Song of the Nibelungs*]. Far from being a naïve expression of catastrophism, however, this book reveals ironic detachment in the excessive pathos of certain passages and the humour and sceptical self-correction of others. Concerns over nuclear testing in the Pacific combine with uneasiness regarding the seemingly effortless suppression of memories of past misdemeanours in the traumatic nightmare visions of Günter Eich's influential radio play *Träume* [*Dreams*] (1950),[17] and prompt him to ask in *Die Stunde des Huflattichs* [*The Day of the Coltsfoot*] (1956)[18] whether man may not lose his position as the pride of creation. The apocalyptic anxieties reflected in the metaphysical poems in Erich Fried's volume *Warngedichte* [*Warning Poems*] (1964)[19] were primarily motivated by the dangers of atomic war, which became a frighteningly real possibility in the Cuban missile crisis of 1962. Here, as in much of his work, Fried attacks apathy, alienation and cynicism. The triumph of consumerist materialism over spiritual needs, the revival of militarism and reservations regarding the growing technological euphoria in the sixties are reflected in Hans Magnus Enzensberger's poems, for instance 'ich, der präsident und die biber' ['I, the president and the beavers'], 'isotop' ['isotope'], and 'das ende der eulen' ['the end of owls'] in the volume *Landessprache* [*Language of the Country*] (1960),[20] and 'doomsday', 'blindenschrift' ['braille'], 'nänie auf den apfel' ['elegy for the apple'] and 'weiterung' ['complication'] in *Blindenschrift* (1964).[21] Enzensberger draws attention to the threat of extinction for many species, including mankind, which is enveloped in its own dangerous technology.

The third phase of catastrophist writing was precipitated by the series of 'doomsday' books which appeared and were avidly read from the early 1970s on. At first these were translations of US authors, for instance Rachel Carson's classic study of the impact of pesticides and weedkillers *Silent Spring* (first published in Germany in 1962, but rediscovered ten years later), and the epoch-making first Club of Rome report *The Limits to Growth* on economic growth and its implications for the consumption of resources, pollution and climatic change (originally published in 1971; in German translation 1972). These American texts were soon followed by German books such as Herbert Gruhl's *Ein Planet wird geplündert* [*The Plundered Planet*] (1975), Rudolf Bahro's *Die Alternative* [*The Alternative*] (1977), and Robert Jungk's *Der Atomstaat* [*The Atomic State*] (1977). Reinhold Grimm was one of the first to examine the new phase of catastrophist writing in 1982.[22] He identified sinking ships and the ice age as key motifs in the literature of the previous decade. These images of disaster, of both individual and planetary annihilation, came to the fore in Tankred Dorst's play *Eiszeit* [*Ice Age*] (1972),[23] and formed a standard repertoire in West Germany and the GDR, Switzerland and Austria in the second half of the decade, in the works of major and minor writers alike, in film and the debate in literary supplements.[24] Grimm notes their ecological significance, but suggests that they do not refer primarily to man's physical destruction of the planet, but rather stand for societal and interpersonal alienation. Societal alienation is particularly important in GDR writing – Grimm refers for instance to Christa Wolf's *Kein Ort. Nirgends* [*No Place on Earth*] (1979). In West Germany, on the other hand, Grimm argues the primary reference is to individual coldness and iciness in interpersonal relationships.

One of Grimm's key examples is Enzensberger's long poem *Der*

Untergang der Titanic [The Sinking of the Titanic] (1978).[25] However, this sophisticated work represents in reality a stance of urbane irony and detachment from doomsday predictions. The sinking of the *Titanic*, the largest passenger ship in the world, deemed unsinkable, that epitome of progress and man's control over nature, on her maiden voyage in April 1912, with a loss of over 1,500 lives, provides a powerful image for the end of faith in technology. But through its association with Castro's Cuba it also stands for the end of Enzensberger's faith in the ability of revolutionary socialism to change society, indeed, in any rational planning for the future of mankind. For, in a poem Enzensberger purports to have written in Havana in 1969, the sinking of the *Titanic* stood for the downfall of bourgeois society at the hand of the proletarian iceberg. The manuscript of this real or imaginary poem has been lost. Now, less than ten years later, with capitalism patently flourishing, the *Titanic*'s fate stands for the end of his apocalyptic belief in revolutionary renewal, and the iceberg becomes a symbol of his dawning realization that life goes on after each disaster.

Enzensberger's much-quoted 'Critique of political ecology', published in the journal *Kursbuch [Timetable]* in 1973,[26] had shown him *au fait* with the arguments of *The Limits to Growth*, but already sceptical of catastrophist predictions, confirming his belief in socialism and man's control over his future. The general decline of ecosocialism and faith in history as linear progress throughout the seventies is reflected in *Der Untergang der Titanic*, where Enzensberger has shifted, despite the wintry sombreness of the scene in Berlin in 1977, to a stance of humour, cheerful defiance, and paradoxical encouragement to the reader to remain alert and shake off melancholy. Loss of utopian hope is presented in an at times confusing, but engaging and fascinating blend of tragedy and satire. Enzensberger is simultaneously rejecting the euphoria of the sixties for technology, and the protest movement against technological progress, with its doomsday predictions. This did not prevent the title of his epic poem, echoing Spengler's *Untergang des Abendlandes [Decline and Fall of the Western World]*, from seeming to encapsulate the new pessimistic spirit of the age. The complexity of the text, its shifts between seriousness and irony, and the subtitle *Eine Komödie [A Comedy]* led to its misinterpretation (for instance by Michael Schneider) as an expression of cynical nihilism.

The clearest evidence of a taste for catastrophism in the GDR in the 1970s is to be found in the cruel nihilist plays of Heiner Müller and the poems of Günter Kunert, both of which contain powerful images for the loss of a perspective of social and technological progress.[27] Kunert's scepticism regarding technology and progress, which goes back to the early sixties, deepened to a view that all pursuit of utopian goals is a delusion in the poems of *Unterwegs nach Utopia* [*En Route to Utopia*] (1977), and, unaffected by his move to the West in 1979, found expression in black pessimism in the poetry and prose of *Abtötungsverfahren* [*Deadening Processes*] (1980), *Verspätete Monologe* [*Belated Monologues*] (1981) and *Stilleben* [*Still Life*] (1983). The poem 'Unterwegs nach Utopia II' ['En route to utopia II'] (1977) illustrates the fusion of social, political and ecological dimensions:

> Auf der Flucht
> vor dem Beton
> geht es zu
> wie im Märchen: Wo du
> auch ankommst
> er erwartet dich
> grau und gründlich
>
> Auf der Flucht findest du
> vielleicht
> einen grünen Fleck
> am Ende
> und stürzest selig
> in die Halme
> aus gefärbtem Glas.[28]

[On the run/ from concrete/ what happens/ is just as in the old tale:/ Wherever you end up/ it's there waiting for you/ grey and thorough// On the run you find/ perhaps/ a green spot/ in the end/ and drop, blissful/ into the blades/ of coloured glass.]

The free flight of hope is dashed, and the transformation of the natural environment into concrete and coloured glass provides, beyond its literal meaning, images for the bureaucratic socialism of the GDR. Kunert's reliance on a stock of familiar symbols allows his critique of contemporary circumstances to be understood individually and collectively, and gives it a timeless dimension. In further poems he indicates an inexorable worsening of the world situation with apocalyptic images of darkness, ruin and decay, being buried, frozen, and turned into stone.

The danger of Europe becoming a theatre of nuclear war in the first half of the eighties gave a new sense of urgency to writing in the west. The year 1983, which shattered illusions about national security and the efficacy of democratic structures (over two-thirds of the population were against stationing the missiles in Germany), saw the burst of catastrophist prose writing referred to at the outset. Nuclear catastrophe reached the stage a little later with plays such as Herbert Achternbusch's *Sintflut* [*Flood*] (1985) and Harald Mueller's *Totenfloß* ['Raft of the dead'] (1986), the deeply pessimistic story of three men and a woman who have survived the nuclear inferno crippled and dying of radiation sickness, and their quest for a land free of contamination and destruction. Mueller's gruesome portrayal of their slow fight against death is alleviated by the highly imaginative dialogue, in which teenage slang is juxtaposed with scientific terminology.[29] Fears of the dangers of technology were given new impetus by the meltdown at the Ukrainian nuclear power station in Chernobyl in April 1986, an event reflected in the writing of Germans in East and West as different as Adolf Muschg, Gudrun Pausewang, Hans-Joachim Schädlich, Gabriele Wohmann and Christa Wolf.

The most important genre in the literature of ecological catastrophism in the eighties is undoubtedly the novel, followed by shorter prose pieces, poems, radio and stage plays. An obvious distinction to be made is that between works extrapolating current economic, technological and social trends into the future, such as Herbert W. Franke's novel *Endzeit* [*Last Days*] (1985),[30] which describes the disastrous climatic consequences of our energy consumption and calls for a change of consciousness and a new, holistic, more responsible way of life, and the much larger number of works depicting nuclear war and its aftermath. Some of the latter attempt a realistic depiction of the consequences of a nuclear double-strike. Matthias Horx, writing in 1982 for the generation of twenty- to thirty-year-olds in the alternative scene, sets his novel *Es geht voran* (subtitled *Ein Unfall-Roman* [*An Accident-Novel*])[31] in Frankfurt in the year 1989. Even before nuclear war breaks out, he describes a society in dissolution, with street gangs and looting, and a population divided into two separate groups: the Synergists, high-tech industrialists who continue to plan for expansion and progress into outer space, and their opponents, the Transformators, Greens preparing individuals to survive the nuclear inferno and restart society thereafter. In his second novel, *Glückliche Reise* [*Happy Journey*] (1983),[32] Horx, inspired by the books of Rudolf Bahro and other green thinkers, develops the model of a new, decentralized society of communes leading a bio-evolutionary lifestyle. However, the cycle of industrialization and militarism repeats itself, and a second catastrophe is needed before a truly promising basis for the new society is given. Volker Lilienthal detects 'a touch of the Romanticism of wilderness and survival training' in not a few catastrophist novels in the early eighties, including those of Horx.[33] Depictions of post-nuclear life with real-istic tips for survival are, even disregarding their implausibility, open to the accusation of actually increasing public acceptance of the idea of nuclear war.

Among the most optimistic works calling for change before it is too late are Gudrun Pausewang's *Die letzten Kinder von Schewenborn* (1983) and *Die Wolke* (1987).[34] The first of the two books deals with the consequences of nuclear war in Germany, concentrating on the devastating effect on one family. Pausewang's gruesome detail is aimed at consciousness-raising in children and their parents. *Die Wolke*, the gripping story of a teenage girl, seeks to shake German readers out of believing nuclear accidents only happen in other countries with less strict safety regulations than in the Federal Republic. A meltdown in the Grafenrheinfeld reactor near Schweinfurt in Hesse in the 1990s leads to 18,000 deaths and the contamination of large areas of the country. Pausewang confronts her young readers with graphic descriptions of the collapse of public order, brutally selfish panic reactions, the death of a young boy who is run over, the effects of radiation sickness, and subse-quent discrimination against the surviving victims of the fall-out. Despite a degree of over-simplification in the presentation of chil-dren and adults, and some heroization, Pausewang's books must be regarded as successful examples of environmental engagement in literature. The case is not so clear for the novel *ENDE. Tagebuch aus dem 3. Weltkrieg*[35] by the Frankfurt journalist Anton-Andreas Guha. This diary of a survivor of nuclear double-strike narrates the final weeks of global destruction. Suffering, inhumanity and death culminate in the suicide of the narrator. The author's extensive factual knowledge permits him to illustrate all the major arguments of the peace movement. However, as critics have pointed out, Guha clearly intends to instil a healthy fear of nuclear war in his readers, but undermines hope of change by his fatalistic biologism. For, like Horstmann, he appears to regard mankind as an error of evolution, and his brutal realism is calculated to provoke resignation among readers rather than stimulating them to opposition.

Any assessment of achievements in the genre of catastrophism would be incomplete without mention of the more complex and ambitious works of Günter Grass and Christa Wolf. Despite their respective West and East German perspectives throughout the seventies and eighties, these two authors shared fundamental ecological concerns, and expressed them in major works on the theme of global catastrophe. In the following I ask first what ecological concerns led to Grass's preoccupation with catastrophism and how successful he has been in integrating his ecological message within literary structures, before concluding with some remarks on the work of Christa Wolf.

The roots of Grass's environmental concern may be located, according to Irmgard Hunt, author of an informative survey of ecological issues in Grass's work,[36] in the very early seventies, and in his anti-capitalist world-view. Grass's speeches in the seventies show growing indignation over misery and injustice in the Third World and the discrepancy between technological and social progress. Technology, he argues increasingly pessimistically, has served more than anything else to facilitate man's destructive urges.[37] In 1975 he experienced at first hand the chaotic slums of India, and later in the decade conditions in the Far East and Africa. Recognition of the seriousness of the global ecological situation was underlined by reading the Club of Rome reports. The first novel to reflect Grass's ecological concerns and present a vision of world catastrophe was *Der Butt* [*The Flounder*] (1977). At the centre of the book is an alternative, feminist version of Runge's tale of the 'Fisherman and His Wife'. Here it is the fisherman whose wishes are insatiable, not his wife, and the story ends not with the couple back in the hut where they started out, but with global destruction:

He wants to be unconquerable in war. He wants to build, traverse, inhabit bridges across the widest river, houses and towers reaching to the clouds, fast carriages drawn neither by oxen nor horses, ships that swim under water. He wants to attain goals, to rule the world, to subjugate nature, to rise above the earth ... And when at the end the fisherman ... wants to rise up to the stars ... all the splendour, the towers, the bridges, the flying machines collapse, the dikes burst, drought parches, sand-storms devastate, the mountains spew fire, the old earth quakes, and in quaking shakes off the man's rule. And cold blasts usher in the next all-covering ice age.[38]

The passage quoted contains in a nutshell all of Grass's ecological concerns and his vision of world catastrophe. Responsible for the latter are restless dissatisfaction, boundless material greed, the squandering of resources and an obsession with actually constructing any device which has been invented, while blaming others for the consequences, whether they be women, men, other races or ideologies.

Kopfgeburten oder Die Deutschen sterben aus [*Headbirths*],[39] a hybrid work published in 1980 which combines the narrative of a film script and the account of a recent tour in the Far East for the Goethe Institute with commentaries on the political situation in Germany, is centrally concerned with issues such as the significance of over-population and the growing misery of the Third World for the future of mankind, and the social and political implications of expanding nuclear energy, but addresses them indirectly, ironically and through imaginative images, and in such a complex narrative framework that the author's standpoint is often far from clear. In the following few years these issues were to merge with concern for the survival of mankind in the face of the nuclear confrontation between the superpowers. Grass's speeches begin to adopt an apocalyptic tone. In 1982 he refers to the Club of Rome reports as 'our down-to-earth Revelation' and 'the apocalypse as balance sheet', paints a truly grim picture of the future, and concludes: 'The destruction of mankind in a variety of ways has already begun.'[40]

Die Rättin, Grass's major ecological novel,[41] had a mixed reception when it came out in March 1986. Three main criticisms were levelled at it: that it was too pessimistic, too didactic, and badly written. Its apocalyptic fable of the end of human civilization brought about by nuclear war, and Grass's diatribes against the stationing of nuclear missiles in Germany seemed anachronistic some two and a half years after the climax of the peace movement, and his alternative vision of the future of mankind as one of creeping destruction of the environment, eventually leading to self-annihilation, looked like yet another version of the doomsday message familiar since *The Limits to Growth*. Had the erstwhile reformer Grass belatedly joined the bandwagon and become a prophet of doom? There were aesthetic criticisms too: the book was seen as too long and too diffuse, as a mere patchwork of statements and narrative fragments. In an unmitigatedly negative review, Marcel Reich-Ranicki called *Die Rättin* 'indigestible' and, unable to resist the pun, 'a catastrophic book'.[42] Grass had, however, himself already used the latter epithet in an entirely positive sense, describing his book as 'a catastrophic book in a catastrophic time'.[43]

Since *Die Rättin* is the subject of another chapter in this volume,[44] it will suffice here to say that this may be an alarmist book, but it is clearly not an example of either the 'stirring up of fears and indulgence in black humour' or 'the postmodern delight in catastrophes' of which Glaser writes.[45] It is a deeply pessimistic work, yet not a fatalist one. For, unlike Guha's book on nuclear war, *Die Rättin* does not prophesy impending doom. In the dispute between the rat and the narrator which forms the core of the novel, Grass leaves open the question of whether the apocalypse is reality or merely a figment of the rat's imagination. Though the disaster has (at least in the view of the rat) already taken place at the outset, the narrator refuses to accept that it has actually happened or must indeed happen. In an interview after publication of *Die Rättin* Grass accepted the description of his role as a writer as that of Cassandra-like 'conjuror-up of catastrophe', and 'literary shock therapy' as a formulation of the intention of *Die Rättin*.[46]

Die Rättin is, according to Wolfgang Ignée, 'Grass's most serious, most radical, most un-fun book'.[47] As an ecological statement *Die Rättin*, like *Kopfgeburten*, is despite this flawed by ambiguities. However, unambiguous eco-evangelism is probably incompatible with the multidimensionality, ambiguity and even obscurity from which works of high literature derive much of their fascination. Grass's novels do not set out to give a balanced assessment of the political or ecological situation, nor do they attempt to provide practical solutions to problems.[48] Grass revealed in a speech in Finland in 1981 that he had been studying the Revelation of St John with a view to his own literary formulation of the likely fate of mankind. To avoid the pitfalls of triviality (he speaks of 'doomsday trash') a modern writer must treat the old myths and the new with sarcasm and irony.[49] His delightful satirical exposure of public humbug in the face of acid rain and dying forests in the novel is undoubtedly one of the strengths of this adaptation of the genre of catastrophism.

It has also been argued that the author's undeniable narrative and stylistic bravura detract from the impact of his message. 'In this entertaining novel, doesn't the end of the world simply become a good read?', Volker Lilienthal asks.[50] Similar accusations have been levelled at Günter Kunert by Michael Schneider, who questions the integrity of 'the despairingly beautiful descending cadences of Kunert's poetry'. By dint of formal mastery this 'poet of sadness' has become a misanthropic 'versifier of chronic depression' and his poetry 'black kitsch'.[51] This ambivalence would appear inherent in the best ecological literature, for instance in the writing of Arno Schmidt,[52] Hans Magnus Enzensberger and Günter Kunert, as well as Günter Grass.

The underlying tension between the complex aesthetic demands of poetic imagery and the persuasive clarity of the ecological message is also present in the work of Christa Wolf, though in a different way. Wolf shared Grass's fundamental criticisms of modern society and his pessimism regarding the future. In a speech in 1981 she called the atom bomb the logical end-product of our civilization, which revealed itself as 'sick, probably insane, possibly dying'.[53] Unlike Grass, however, her two main contributions to catastrophism, *Kassandra* (1983) and *Störfall* (1987),[54] set out to examine the psychological and social roots of aggression and technological self-destruction.[55] Wolf's cultural feminist interpretation of the development of western civilization in *Kassandra* roused international interest. *Kassandra* tells the story of the Trojan War from the perspective of the prophetess whose insight into approaching disaster has made her name synonymous with doom-laden pessimism. It combines this narrative with theoretical reflections by presenting the mythological tale alongside four 'essays', called 'Prerequisites of a story', actually lectures on literature and her own writing given at the University of Frankfurt in 1982. These incorporate travelogue, diary entries and a long letter, revealing the personal and universal experiences which accompanied the creative process, and juxtaposing mythological material with reflections on the contemporary world. The *Kassandra* complex thus constitutes a 'narrative network' combining the essayistic and narrative elements Grass sought to integrate in his novels.

Wolf portrays the present as a watershed, with apocalyptic, but equally utopian potential, and explores the individual and social factors behind war, locating them in particular in ideology, self-seeking and male chauvinism. Only a historic cultural shift can save the world. Wolf provides, both in Kassandra's perception and self-realization and in the cave community outside the walls of Troy, images of alternative values which could avert global destruction and offer the potential of a post-patriarchal society. *Kassandra*'s success in expressing contemporary fears and hopes is reflected in its achievement of cult status in both German states, and selling nearly half a million copies in the West.

Störfall (1987), Wolf's best-selling response to the Chernobyl disaster, a book which takes up many of the ideas discussed by Grass in his speeches and in *Die Rättin*, presents a much bleaker view of the future of mankind. The accident in the Ukrainian power station becomes a paradigm for the threats created by modern science, and the reaction of the authorities (the Soviets only admitted it had taken place two days later, and governments in east and west played down the consequences for public health) serves as the ultimate disillusionment of her belief in social progress. She opens with a vision of the end of the world: 'On a day about which I cannot write in the present tense, the cherry trees will have been in blossom.'[56] Wolf's uncompromising apocalyptic tone in this book is paralleled in GDR literature only by Kunert and Müller. She discusses the irrational depths of the psyche and the terrifying nature of repressed instincts, and identifies an inborn tendency towards violence and destruction, the evolution of a restless, hyperactive human brain, gender-specific qualities, and socialization as factors in our obsessive preoccupation with dangerous technologies.

Störfall was given a very positive reception in the GDR, where it fulfilled an important function in opening up a hitherto taboo subject to public debate. However, it was greeted in the West as disappointing, embarrassingly unfinished, and lacking in originality and artistic refinement. The most obvious explanation for weaknesses, such as the uneasy match of Wolf's spontaneous emotional response to the news of the accident with reflections on the ultimate causes of our propensity to engage in dangerous and globally destructive technologies, is that *Störfall* was written hastily (between June and September 1986), because Wolf was conscious that her privileged position would enable her to make a significant contribution to public debate on this very important topic in the GDR.

The awkward dichotomy of aesthetics and ecological message has thus only been partially resolved in the individual works examined here, and the literature of catastrophism has revealed itself as playing a highly ambiguous role in ecological debate. Catastrophism undoubtedly reflects an undercurrent of pessimism and irrationalism in German culture. With some justification, Wolf and Grass have been accused of lapsing into anti-modernist diatribes, dismissing the achievements of scientists and engineers as part of a trend towards corruption and decadence. The aesthetically most satisfying works in the catastrophic genre, for instance those by Arno Schmidt, Hans Magnus Enzensberger, Günter Kunert, Günter Grass and Christa Wolf, are often ambiguous as ecological statements, and paradoxically tend to subvert doomsday predictions with ironic detachment.

Nonetheless, catastrophist literature has acted as a corrective in a predominantly anthropocentric and technocratic culture, disseminating and elaborating green ideas, and sketching out social alternatives. Without always compromising literary quality unduly, it has played a significant part in illustrating and problematizing the post-industrial values of the new social movements,[57] in confirming the fears and aspirations of readers, and in promoting a change in consciousness.

Notes

[1] Hermann Glaser, *Kulturgeschichte der Bundesrepublik Deutschland 1968–89* (Munich, 1989), 216.

[2] 'Apokalypse, Politik als Psychose und die Lebemänner des Untergangs' and 'Die Intellektuellen und der Katastrophismus: Krise oder Wende der deutschen Aufklärer?', reprinted in Michael Schneider, *Nur tote Fische schwimmen mit dem Strom. Essays, Aphorismen, Polemiken* (Cologne, 1984), 34–133.

[3] Gunter E. Grimm, Werner Faulstich and Peter Kuon (eds.), *Apokalypse. Weltuntergangsvisionen in der Literatur des 20. Jahrhunderts* (Frankfurt am Main, 1986).

[4] Klaus Vondung, *Die Apokalypse in Deutschland* (Munich, 1988).

5. Jost Hermand, *Grüne Utopien in Deutschland. Zur Geschichte des ökologischen Bewußtseins* (Frankfurt am Main, 1991). See the sections 'Ökologische Warnungen in der Nachkriegszeit', 'Vom "Wohlstand für alle" zum "Bericht des Club of Rome"', 'Von der Anti-Atomkraft- und Friedensbewegung zur Partei der Grünen' and 'Reaktionen auf neue Hiobsbotschaften in den achtziger Jahren'.

6. Reinhold Grimm, '"The ice age cometh": a motif in modern German literature', in Siegfried Mews (ed.), *'The Fisherman and His Wife'. Günter Grass's 'The Flounder' in Critical Perspective* (New York, 1983), 1–17. Previously published in German: 'Eiszeit und Untergang: Zu einem Motivkomplex in der deutschen Gegenwartsliteratur', *Monatshefte*, 73, 2 (1982), 155–86.

7. Paul Konrad Kurz, *Apokalyptische Zeit. Zur Literatur der mittleren 80er Jahre* (Frankfurt am Main, 1987).

8. Volker Lilienthal, 'Irrlichter aus dem Dunkel der Zukunft. Zur neueren deutschen Katastrophenliteratur', in Helmut Kreuzer (ed.), *Pluralismus und Postmodernismus. Zur Literatur- und Kulturgeschichte der achtziger Jahre*, 2nd edn. (Frankfurt am Main, 1991), 190–224.

9. Principally articles on Enzensberger, Kunert, Grass and Wolf. References are given below.

10. Ulrich Horstmann, *Das Untier. Konturen einer Philosophie der Menschenflucht* (Frankfurt am Main, 1985).

11. Hoimar von Ditfurth, *So laßt uns denn ein Apfelbäumchen pflanzen. Es ist soweit* (Munich, 1988).

12. Vondung, *Die Apokalypse in Deutschland*, 8f.

13. Schneider, *Nur tote Fische schwimmen mit dem Strom*, 107–24.

14. Vondung, *Die Apokalypse in Deutschland*, 12.

15. Arno Schmidt, *'Brand's Haide'. Zwei Erzählungen*, reprint of the 1st edn. (Frankfurt am Main, 1985).

16. Arno Schmidt, *Kaff auch Mare Crisium*, reprint of the 1st edn. (Frankfurt am Main, 1985).

17. Günter Eich, *Gesammelte Werke*, 2 (*Die Hörspiele I*), ed. Heinz Schwitzke (Frankfurt am Main, 1973), 287–327.

18. Günter Eich, *Gesammelte Werke*, 3 (*Die Hörspiele II*), ed. Heinz Schwitzke (Frankfurt am Main, 1973), 921–55. A later version (1959) is also included in this edition of Eich's works, ibid., 1225–69.

19. Erich Fried, *Warngedichte* (Frankfurt am Main, 1980).

20. Hans Magnus Enzensberger, *Landessprache. Gedichte* (Frankfurt am Main, 1981).

21. Hans Magnus Enzensberger, *Blindenschrift* (Frankfurt am Main, 1967).

22. Grimm, '"The ice age cometh"'.

23. Tankred Dorst, *Eiszeit* (Frankfurt am Main, 1973).

24. See also Thomas Koebner, 'Endzeit bei Hans Magnus Enzensberger,

Günter Kunert und Heiner Müller', in *Unbehauste. Zur deutschen Literatur in der Weimarer Republik, im Exil und in der Nachkriegszeit* (Munich 1992), 368–97. Koebner identifies four fields of metaphors in the literature of catastrophe in the eighties: explosion (destroying the old, liberating from the evils of history), shipwreck, the associated metaphors of rubble, refuse, ruins and ash, and, finally, coldness and ice. These are central elements in a literary discourse on human hybris and punishment.

25. Hans Magnus Enzensberger, *Der Untergang der Titanic. Eine Komödie* (Frankfurt am Main, 1978). Translated into English by the author and published as *The Sinking of the Titanic: A Poem* (Manchester, 1981). The following comments are based on Joseph Kiermeier-Debre, '"Diese Geschichte vom untergehenden Schiff, das ein Schiff und kein Schiff ist". Hans Magnus Enzensbergers Komödie vom "Untergang des Untergangs der Titanic"', in Grimm et al., *Apokalypse*, 222–45; and Moray McGowan, '"Das Dinner geht weiter": some reflections on Hans Magnus Enzensberger and cultural pessimism', in Heinrich Siefken and J. H. Reid (eds.), *Lektüre – ein anarchischer Akt* (Nottingham, 1990), 5–17.

26. Hans Magnus Enzensberger, 'Ökologie und Politik oder Die Zukunft der Industrialisierung', *Kursbuch 33* (1973), 1–42. Translated as 'A critique of political ecology', in *Dreamers of the Absolute: Essays on Politics, Crime and Culture* (London, 1988), 253–95.

27. Kunert's work is discussed by Thomas Koebner in 'Endzeit bei Hans Magnus Enzensberger, Günter Kunert und Heiner Müller', and in 'Apokalypse trotz Sozialismus. Anmerkungen zu neuen Werken von Günter Kunert und Heiner Müller', in Grimm et al., *Apokalypse*, 268–93. See also Walter Hinderer, 'Arbeit an der Gegenwart. Apokalyptische Signale in Günter Kunerts Lyrik von 1966–1990', in Lothar Jordan and Winfried Woesler (eds.), *Lyriker Treffen Münster. Gedichte und Aufsätze 1987–1989–1991* (Bielefeld, 1993), 456–77, Grimm, '"The Ice Age Cometh"', and Michael Schneider, 'Wie depressiv sind unsere Poeten? Über Günter Kunert und ZEITgenossen', in *Nur tote Fische schwimmen mit dem Strom*, 141–59.

28. Günter Kunert, *Unterwegs nach Utopia. Gedichte* (Munich and Vienna, 1977), 76.

29. Harald Mueller, *Totenfloß*, *Spectaculum* 43 (Frankfurt am Main, 1986), 77–125.

30. Herbert W. Franke, *Endzeit* (Frankfurt am Main, 1985).

31. Matthias Horx, *Es geht voran. Ernstfall-Roman* (Berlin, 1982).

32. Matthias Horx, *Glückliche Reise. Roman* (Berlin, 1983).

33. Lilienthal, 'Irrlichter aus dem Dunkel der Zukunft', 197.

34. Gudrun Pausewang, *Die letzten Kinder von Schewenborn. Oder . . . sieht so unsere Zukunft aus? Erzählung* (Ravensburg, 1985), and *Die Wolke. Jetzt werden wir nicht mehr sagen können, wir hätten von nichts*

gewußt (Ravensburg, 1989).

[35.] Anton-Andreas Guha, *ENDE. Tagebuch aus dem 3. Weltkrieg* (Kronberg am Taunus, 1983).

[36.] Irmgard Hunt, 'Vom Märchenwald zum toten Wald: ökologische Bewußtmachung aus global-ökonomischer Bewußtheit. Eine Übersicht über das Grass-Werk der siebziger und achtziger Jahre', in Gerd Labroisse and Dick van Steckelenburg (eds.), *Günter Grass: Ein europäischer Autor?*, Amsterdamer Beiträge zur neueren Germanistik, 35 (Amsterdam and Atlanta, 1992), 141–68. Secondary literature on the catastrophist aspects of Günter Grass's work includes Sigrid Mayer, 'Zwischen Utopie und Apokalypse. Der Schriftsteller als "Seher" im neueren Werk von Günter Grass', in *Literarische Tradition heute. Deutschsprachige Gegenwartsliteratur in ihrem Verhältnis zur Tradition*, Amsterdamer Beiträge zur neueren Germanistik, 24 (Amsterdam, 1988), 79–116; Wolfgang Ignée, 'Apokalypse als Ergebnis eines Geschäftsberichtes. Günter Grass' Roman Die Rättin', in Grimm et al., *Apokalypse*, 384–401; Thomas W. Kniesche, *Die Genealogie der Post-Apokalypse. Günter Grass' Die Rättin* (Vienna, 1991); and Volker Neuhaus, 'Günter Grass' *Die Rättin* und die jüdisch-christliche Gattung der Apokalypse', in Gerd Labroisse and Dick von Stecklenburg (eds.), *Günter Grass: Ein europäischer Autor?*, 123–39.

[37.] See Günter Grass, *Werkausgabe in zehn Bänden*, edited by Volker Neuhaus, 9, *Essays, Reden, Briefe, Kommentare* (Darmstadt and Neuwied, 1987), 674f. and 890.

[38.] Günter Grass, *The Flounder*, translated by Ralph Manheim (San Diego, New York and London, 1978), 349.

[39.] Günter Grass, *Kopfgeburten oder Die Deutschen sterben aus* (Darmstadt, 1988).

[40.] Günter Grass, *On Writing and Politics 1967–1983*, translated by Ralph Manheim (San Diego, New York and London, 1985), 137.

[41.] See Heinz Ludwig Arnold's analysis of the reception of the book, in 'Literaturkritik: Hinrichtungs- oder Erkenntnisinstrument. Günter Grass' Rättin und das Feuilleton', *L '80*, 39 (1986), 115–26.

[42.] Ibid., 116f.

[43.] 'Mir träumte, ich müßte Abschied nehmen', television interview with Beate Pinkerneil, recorded 3 March, and broadcast on ZDF, 24 March, 1986; in Grass, *Werkausgabe*, 10, *Gespräche mit Günter Grass*, 360.

[44.] See ch. 10.

[45.] Glaser, *Kulturgeschichte der Bundesrepublik Deutschland*, 218.

[46.] Grass, *Werkausgabe*, 10, *Gespräche mit Günter Grass*, 352 and 360.

[47.] Ignée, 'Apokalypse als Ergebnis eines Geschäftsberichtes', 397.

[48.] Grass's political views are more explicit in speeches such as 'Orwells Jahrzehnt 1', where he discusses with enthusiasm ideas in Willy Brandt's report for the North–South Commission (*Das Überleben sichern*), such as 'a new world-wide economic order', 'reduced consumption', 'end of the exploitative economy' and 'world-wide disarmament'.

[49.] Grass, *Werkausgabe*, 9, 793.

[50.] Lilienthal, 'Irrlichter aus dem Dunkel der Zukunft', 209.

[51.] Schneider, 'Wie depressiv sind unsere Poeten?', 148–50 and 156.

[52.] Jörg Drews writes of Schmidt: 'Schmidt has such fun with his own black humour and bizarre inventions that light-hearted amusement overwhelms the terror.' See '"Wer noch leben will, der beeile sich". Weltuntergangsphantasien bei Arno Schmidt (1949–1959)', in Grimm et al., *Apokalypse*, 14–34, here 21. See also 31 and *passim*.

[53.] *Berliner Begegnung zur Friedensförderung. Protokolle des Schriftstellertreffens am 13./14. Dezember, 1981* (Darmstadt and Neuwied, 1982), 117.

[54.] Christa Wolf, *Kassandra. Vier Vorlesungen. Eine Erzählung* (Berlin and Weimar, 1983), and *Störfall. Nachrichten eines Tages* (Darmstadt and Neuwied, 1987).

[55.] From the very considerable body of secondary literature on Wolf I am indebted to Jacquie Hope, *Green Trends in East Germany: Critiques of Modern Industrial Society in GDR Literature*, D.Phil. thesis, Oxford, 1992, J. H. Reid, *Writing without Taboos: The New East German Literature* (New York, Oxford and Munich, 1990), and Anna Kuhn, *Christa Wolf's Utopian Vision: From Marxism to Feminism* (Cambridge, 1988).

[56.] Christa Wolf, *Accident: A Day's News*, translated by Heike Schwarzbauer and Rick Takvorian (New York, 1989), 3.

[57.] See Stephen Brockmann, Julia Hell and Reinhilde Wiegmann, 'The Greens: images of survival in the early 1980s', in Reinhold Grimm and Jost Hermand (eds.), *From the Greeks to the Greens: Images of a Simple Life* (Madison and London, 1989), 127–44.

10 • Ecological Disaster as a Narrative Precept in Günter Grass's *The Rat*

Johann Siemon

From the seventies onwards, Günter Grass became increasingly concerned with environmental issues in his work, a concern reflected in his attempts to communicate the importance of rethinking our attitude towards consumption and technical progress. In time, however, his belief in the possibility of change was replaced by an ever-gloomier outlook which finally resulted in a catastrophic pessimism. In his novel *Die Rättin* [*The Rat*], published in 1986, he envisaged a global catastrophe and brought the dialectic of mutual military threat and international environmental destruction to its logical conclusion: human self-annihilation.[1]

This global destructive force influences the thematic as well as narrative structures of the book and infects the writing process itself. In consequence it leads to the destruction of traditional thinking and narration. Using the example of *Die Rättin* I want to discuss the implications of fictional self-annihilation for narration by concentrating on the following three points:

1. The end of (human) history
2. The disappearance of the narrator
3. The end of narration.

In so doing I wish to adopt Kniesche's five categories of narration, and integrate them as far as possible into my own categories.[2] Kniesche points out that the different narrative tracks constitute a web of referentiality and allusion that is dominated by the central symbol[3] of the rat, which gives the book its critical dynamic. In this way, the animal that in the main has previously tended to be interpreted in negative terms, and has fostered connotations of repulsiveness and disgust, now becomes a positive force in the text.[4] The rat assumes the function of narrator, critic, friend and pessimistic prophet of man, as well as being a survival specialist who in the end will become the dominating post-human form of existence.[5]

1. The end of (human) history

The hypostasized failure of enlightenment forms the intellectual background for the description and analysis of negative historical developments and their sudden end. In an interview with Beate Pinkerneil, Grass describes uncritical belief in the progress of the technocratic view of the world as a perversion of rationality: 'The identification of rationality with the technical and the possible is for me the dead end of European Enlightenment.'[6]

Human beings are not seen as an autonomous entity, but become a function in the process of communication and production in society. The imposed supremacy of technocratic thinking is accompanied by a dehumanization of all areas of life, which radically calls into question the basic belief in progress that underlies such thinking. The destruction of this concept, however, does imply the rejection of the hierarchic model of Platonic thinking and of the Hegelian concept of history.[7]

We are well acquainted with this form of critical discourse as part of modernism, so there is nothing very original about it. Nevertheless, it does develop a unique perspective as the author does not merely criticize the deplorable state of affairs and the negative developments, but relates the specific destructive elements to their historical and socio-political situation, so that the nuclear catastrophe becomes the inevitable end of progress. Thanks to an understanding of enlightenment that is based on a perverted and superficial concept of group interest, the attempt to enlighten the human race[8] is transformed into a mechanistic process, which – instead of leading man to mental maturity – causes his ruthless degradation to an object and his final destruction. How does Grass – with all this in mind – set up the destruction of history in the text? I think he does it by using three categories which, though chronologically related, are not presented in a chronological order in the text, but rather deconstruct into fragmented, associative facets reminiscent of a *mémoire involontaire* technique. History appears in the text above all as fragmented reference and conceptual quotation. The names of concentration camps that the rat enumerates in her narrative, for example, function as an incantation ritual that refers to the historical mass extermination while emphasizing its inconceivability. But at the same time they form, in the shape of repressed horror, the background of the imminent, ultimate self-destruction and thereby refer to a process of enlightenment which never took place. Not even after the madness of National Socialism did people try to understand what they are capable of, but – and this is where a chance was missed to make a historical break – they took flight into a condition of suppression which allowed them to remain with their emotional and pragmatic dispositions while at the same time making a superficial new democratic start. The narrative strand which focuses on the forger Malskat makes it clear that the cliché of a fresh start in 1945 (variously described in German as the 'Stunde Null' or 'Kahlschlag') actually amounted to silently passing over historical events. The Malskat story implies that, given these assumptions, history could be no more than a suppression or falsification of the past. The painter, whose 'talent to be truly gothic despite poor payment, was fashionable in those days and a reflection of a basic need, namely the general wish for forgery' (*R*, 80), acts as an example for the whole of post-war Germany. The fact that he – together with Adenauer and Ulbricht – is introduced as the 'German triumvirate of forgers' (*R*, 42) makes the point that historical narration in the East and the West is not only based on a lie, but as a consequence of that, will always continue to become one. While Malskat finally comes forward himself and is sentenced for forgery, it remains the 'time of elbow-nudging, of the denazification certificate and of pleasant appearances' for all the rest. 'In the decade of the little innocent who has a clean slate, of the murderer in high office, of the Christian hypocrite on the government bench, nobody wanted to know exactly what had happened' (*R*, 329). In the absence of other values, politics propagates economic growth, while its personnel, including the chancellor, becomes hollow and interchangeable (*R*, 51).[9]

Science with all its 'expert reports, counter-reports, statistics of harmful substances' (*R*, 50) does nothing but legitimize progress. The topic of dying forests exemplifies that established interests have done everything to blur the facts about the actual conditions by using the media to manipulate the perception of reality in the way that is regarded as useful (*R*, 50–7). Or, in the words of Oskar Matzerath: 'People have had enough of the documentary. So much reality is tiring. Nobody believes in facts anyway. Only dreams from the box of tricks produce proper facts. Let's not fool ourselves: truth's name is Donald Duck, and Mickey Mouse is its prophet!' (*R*, 86f.). This is a cynical postmodern attitude that categorically denounces the truth concept and replaces it with the absolute relativism of 'anything goes', and that is wholly concerned with the criterion of commercial success. With the introduction of a critical opposition, however, the narration contradicts this one-dimensional development. The example of the chancellor's children makes it clear that the younger generation could become a potential hope, if they give up the life-style and values of their parents and by doing so acquire a less ideological view of reality. However, in the book this possibility of a critical function in and for society is limited to the minority group of the punks. As members of this movement, the chancellor's children themselves become outsiders and are persecuted by a society which brings them into close relationship with the symbol of the rat. Their attempt to escape from society refers back to the folk- and fairy-tale tradition, and thereby initiates a new interpretation of 'The Pied Piper of Hamelin' and 'Hansel and Gretel'.

What is important at this point is that their rebellion remains a subcultural phenomenon and therefore fails to have an impact on society as a whole. Everything continues to move 'in the wrong direction' (*R*, 229). Intermediate-range missiles[10] are stationed to secure freedom (*R*, 246), and the neutron bomb is supposed to guarantee that all cultural monuments outlive human beings undamaged (*R*, 244). That such thinking is no longer aware of its inherent madness – meaning can only be ascribed to any culture by and through human beings – drastically marks a mental condition that has erased imminent dangers and threats from its consciousness. This way of thinking forgets that history is based on the existence and experience of the individual, and that without such experience, history will cease to be invented and passed on. The dialectic of the optimistic belief in progress and the firm control and domination of the object world leads to a situation where man overestimates his abilities and therefore regards himself as omnipotent and loses his natural fear.[11]

This is a constellation which, hubris-like, provokes the revenge and retaliation of the gods. The nemesis in the text, however, is not triggered by a god but by man himself. The system error which leads to the launching of the nuclear missiles, is, whether through manipulation or chance, self-produced and therefore people's own fault. But the catastrophe in the text, far from being a unique event, is conjured up again and again; like a terrible chorus, like an admonishing ritual, it constitutes the sediment of narration.[12] Paradoxically it is from this ultimate point that the narration derives its justification, as a warning for the narrator's present and its unavoidable absurdity in relation to post-human time. In the moment of catastrophe the sudden alteration of all possible and constructed meaning into absolute meaninglessness takes place: 'A little tired of the historic burden we/ date the end of all history' (*R*, 330).

The author, however, rather than ending with this abrupt destruction of meaning and history, adds an epilogue which allows the gradual disappearance of meaning in the remains and documents of human existence to be revealed. Even the beginning of post-human time is not a new start, but a more or less obvious perpetuation of human behaviour and thinking. Human beings have not disappeared from the surface of the earth altogether: a genetically manipulated rat-man combines characteristics and qualities of both creatures and is therefore supposed to guarantee the survival of the human species. It is, however, above all the human heritage that causes the so-called Watsoncricks to establish a master–servant relationship between themselves and the rats: 'They represent power, but not blind aggression. They have an inborn casual discipline. Without being bitten by guards – a procedure that is unfortunately necessary among rats – they follow their imperceptible order' (*R*, 426). Neither the dominance of the females nor their neutrality and their social democratic behaviour can prevent them from becoming ever more militaristic, so that they finally seize power in the post-human era.

Seventy-five years later they lead élitist and decadent lives: there are occurrences of cannibalism and mass slaughter until the rats revolt and kill the Watsoncricks. Only then is the world 'finally free of humans' (*R*, 504). The human remains, represented by the city of Danzig, disintegrate, and with them the last traces of human history.[13]

2. The disappearance of the narrator

The end of history and the disappearance of the narrator are closely linked within the narrative structures, as both elements are an effect of the death of humanity. But in a strange twist which could reasonably be regarded as a case of inconsistency, albeit one without which the book could never have taken on its present form, the narrator's voice outlives his physical existence. The author not only knows about this contradiction, but actively uses it to undermine the authority of the first-person narrator by replacing the monological with a dialogical structure in order to keep open the question of who is telling the story. An apparently unmotivated metaphor of space is used to make it clear that the status of the first-person narrator changes: very early in the book he is taken out of his familiar environment and becomes isolated in a way that clearly limits his freedom of movement and at the same time points to his estrangement from reality. His movement from a wheelchair (*R*, 34) to a space capsule (*R*, 35) to the Ultimo-observation-satellite (*R*, 160) marks his topographical disappearance from earth. At the same time this distancing metaphor points to another aspect: by widening the gap between himself and observed reality the narrator mimics the so-called objective view of the scientist; in other words, he becomes observer of an experiment. That state of affairs confronts the reader with the subject–object dichotomy, which is now fought over between the first-person narrator and the narrating rat. This fight over the correct narrative perspective and the narrative power is at the same time a vehicle for the critique of the scientific conception of the world that propagates a power-orientated possessiveness. The misunderstood biblical quotation – populate the earth and subdue it[14] – patently results in the appropriation of the object world and an understanding of nature only in terms of exploitable value. It is exactly this structure that dominates the relationship between the rat and the first-person narrator at the beginning of the book. The narrator wants a rat for Christmas in the hope that the rodent will supply him with some inspirational words for a poem on the enlightenment of mankind (*R*, 7). The animal therefore acquires the status of a possession, of an object – albeit for noble motives – to further the narrator's own needs. But the instrumentalization of the rat as a catalyst for the human imagination fails. It turns out that the animal has not only a life but a will of its own, which contradicts, undermines and finally rejects all human claims to power. 'My Christmas rat ... She senses me with her whiskers. She plays with my fears which are convenient for her.

So I talk against her' (*R,* 9). This short quotation already indicates that the relationship between the first-person narrator and the rat will be wholly different from the way man planned it. Instead of the birth of Christ, and his promise of peace and redemption, the Christmas rat is the centre of interest and awakens the subconscious fears that point to the final defeat of the project of enlightenment and the possible end of humanity. Right from the start the narrator is on the defensive, and has to contradict everything the rat symbolizes for him. No objective aloofness and neutrality is ever possible. To avoid the fearful consequences that the rat indicates for the narrator, the actual threat has to be kept unconscious. In consequence, the rat dominates the narrator's dreams: 'Recently the rat has come into my dreams ... my daydreams, my night-dreams are her marked territory' (*R,* 10). The author is able to use the dream motif to break down conceptual dichotomies like subject–object, master–servant, optimism–pessimism, truth–lie, fact–fiction to transform them into a dialogic openness or vagueness without immediately having to transcend them by a Hegelian synthesis or some mystic dispersal of the *principium individuationis.* Instead the discussions and narrations allude to the important enlightenment tradition of the Socratic dialogue. The dispute over the authoritativeness of each narrative is simultaneously an argument about truth, and that relies solely on the question of which argument is the more convincing.

In analogy to a work by Jean Paul,[15] which imagines the speech of the dead Christ from the top of the world to the effect that there is no God, 'the rat speaks out from the mountain of garbage' (*R,* 14) and declares that garbage is the only human legacy. She refutes all the narrator's objections and assurances that humanity still exists with credible and detailed descriptions of the nuclear catastrophe, which the first-person narrator in his space capsule is even able to verify, if only in his dreams: 'The worst predictions confirmed. It's enough to say: through the earth-facing oval window of my capsule everywhere looked terrible, especially Europe; no, everywhere' (*R,* 141). The obvious validity of the rat's description both makes the first-person narrator increasingly uncertain and renders his anti-narrative[16] a stubborn and unreasonable enterprise that is based only on absurd, defiant hope.[17]

Since the narrator can no longer claim to be telling the truth or reflecting reality, he must redefine his function in search of renewed legitimation for his writing. The alternative is to forfeit his power of narration altogether. The definition of his function as writer, namely that 'I want to tell stories, because I can use words to put off the end' (*R*, 16), seems to reflect a mainly subjective interest. But this incessant replacement of the end in the act of writing has of course more general implications, as the main aim of this writing-for-life is a postponement during which humanity could come to its senses.[18] So the annihilation of time is to make room for the desperate hope that change is still possible. Another pattern that functions with the same logic is the perpetual rhetoric of saying goodbye.

In the Pinkerneil interview Grass confirms this interpretation:

> If you take the beginning of the fourth chapter, there is a poem called 'I dreamed I had to say goodbye.' In this poem, through sheer joy of life, everything is enumerated that brings happiness, from the small things to ideas that are part of humanity. That shows that I cling to life and don't know anything better.[19]

But while the narrator sketches a picture of his life by means of enumerating words and phrases evoking certain circumstances and objects, by inventing these lists, he simultaneously indicates the theoretical endlessness of any such enumeration, a point which the narrator underlines by using the rhetoric of saying goodbye repeatedly throughout the book. On a more theoretical level this enumerative mechanism is a justification of narration: since there are innumerable versions of one's own life, there can be no end of story-telling. If the experience of mortality creates the impulse to narrate, the story acquires two functions, at once celebrating and taking leave of life. The first-person narrator represents everyone's right to one's own story, and in his narration he rebels against the possibility that man, through his own fault, should have lost this right. And as the subject–object complex remains in balance until the end, and the question of the narrative authority is never quite answered, this view of the narrator is to some extent supported by the textual strategies. A poem at the beginning of the eleventh chapter stresses this never-ending interpretative circle in which dream and reality dissolve into each other: 'I dreamt of a person/ said the rat that I'm dreaming of./ I talked to him, until he believed,/ he dreamt of me and in the dream he said: the rat/ I'm dreaming of thinks she is dreaming me;/ in this way we are reading each other in mirrors/ and question each other.// Is it possible that both/ the rat and I/ are dreamt of and are a dream/ of yet another kind?' (*R*, 423). In this fantasy of a third kind, be it in the author's head or the reader's, the book ends. The first-person narrator and the rat dissolve, perhaps to make room for the hope that there could be a third way between blind trust and catastrophe, between the self-destructive powers of progress and apocalypse, that humanity could find the right measure.

3. The end of narration

That a narration of the self-destruction of the human race has paradoxical implications is self-evident. I therefore intend to concentrate not on that paradox, but on another, namely the way in which this state of affairs negates the possibilities of narration, while at the same time, in a complex, even desperate manner, continues to support it. With the end of humanity, literature loses its function, since its construction of meaning, values and world that is the basis of all cultural tradition can only take place in the dialectic of production and reception.[20] The author exemplifies this loss of meaning by referring on the one hand to his own work, and on the other to the folk- and fairy-tale tradition. However, he uses these sources not only in a quotational sense, but also, as Genette calls it, as metafictional literature, that is, as hypertexts.[21] He explains this term as follows: 'I call any text hypertext that has been deduced from another text through a simple transformation (we will call this simply transformation) or through an indirect transformation (through imitation).'[22] Which sort of deduction does the author use in our case? Within the categories of Genette the hypertexts we are confronted with here could be called resumption or fictional epilogue.[23] Neither category finishes an unfinished work, as does for example the continuation, but rather they 'go beyond what originally counted as the ending'.[24] In those cases, then, the author does not alter the model, what Genette would call the hypotext, but puts the characters in an up-to-date context, within which they can, according to their role expectation or other requirements, become acting subjects. The character's fate in the presence of the book is, however, not without effect on the updated source. For, as Genette points out: 'there are no innocent transpositions – that is, none which do not in one way or the other modify their hypotexts.'[25] In our case that means that if the hypertexts are destroyed in *Die Rättin*, then this will also influence their hypotexts. In the case of the folk- and fairy-tale tradition, this will even be true for literature as a whole. I intend to look at this complex of problems first in terms of Grass's work, then of fairy-tales.

Stolz has already made it clear that *Die Rättin* may be read as a stock-taking of all aspects of Grass's work up to 1986.[26] It is not possible here to look at these aspects in any greater detail. My observations will therefore be limited to the narrative concerning Oskar Matzerath, the main character in *Die Blechtrommel* [*The Tin Drum*], and that concerning the *New Ilsebil*, the research ship that features in *Der Butt* [*The Flounder*].[27]

Die Blechtrommel, originally published in 1959, is one of the most important novels of the Adenauer era. From the distorted and revealing viewpoint of the dwarfish Oskar Matzerath, it describes German society and life from before the war until the fifties. If, therefore, the Oskar Matzerath story is resumed in *Die Rättin* (as it is), the story must inevitably become an interpretative modification of the fictional world of the Danzig trilogy, and, thanks to its exemplary status, even beyond that. In *Die Rättin*, Oskar appears as an almost sixty-year-old manager of a film company who has specialized in pre- and post-apocalyptic videos. The main attraction of that sort of film seems to lie in the fact that they exclude reality in order to replace it with a stylized apocalyptic thrill. However, the audiences fail to realize that Matzerath's fictions have a tendency to become real: 'Believe me', Oskar tells the narrator, 'tomorrow we will create a reality by which the future, thanks to the constructivity of the media, will lose all its vagueness and accidental nature; whatever will be, it can be' (*R*, 101). The distinction between fact and fiction no longer exists. Reality is such that the media can predict the future. Multifaceted reality no longer guarantees an endless number of possible developments; everything results in 'human end-game programmes' (*R*, 104) instead. Matzerath's films are all about the generally suppressed truth that humanity has no future. Under these conditions reality becomes a simple repetition of media productions, a poor imitation of a film.

The past is just as closed as the future. Oskar Matzerath's return to the area around Danzig to celebrate the 107th birthday of his grandmother is the disappointment of all nostalgic hope. Nothing has remained of the Poland he remembers: Anna Koljaiczek had to move to another area, her potato fields have, 'as many another myth, disappeared under concrete' (*R*, 218). Only the grandmother, this 107-year-old example of a working-class history, this corrective of mainstream history, remains as a promise of security that never existed. When the reality finally catches up with the film and the nuclear catastrophe happens, the beginning and end of Grass's work meet, as the dying Oskar returns under the skirts of his grandmother. Even if Anna Koljaiczek continues to exist as a mummified icon of the rat herds for a while, she does so only as an allegory of annihilation, as her (hi)story has disappeared with her. In this way the author begins the destruction of his own work. But eliminating a self-willed character like Oskar Matzerath is no easy matter (*R*, 394). Suddenly, and against the narrator's will, he turns up again and demonstrates his power over the narrator's mind. Oskar Matzerath once again forces his way into the narrative and obstinately celebrates his sixtieth birthday. Even the nuclear catastrophe cannot be contemplated without this productive genius, especially since it is by no means clear that this apocalyptic scenario, complete with its array of motifs, is not just another work from the Matzerath production line. Neither would this interpretation necessarily be comforting, for if that were the case, this story would predict the future, which means, if one shares Oskar Matzerath's totally deterministic view, that the end is nevertheless inevitable.

Der Butt (originally published in 1977), was in one respect, and put simply, a chauvinistic version of the fairy-tale *The Fisherman and His Wife*, which questions the relationship between the sexes and discusses the contribution of women to history. As Barbara Garde puts it, the book was 'the high point of a widespread public debate about the new German women's movement'.[28] In another respect, the book uses historical developments and contexts to establish a critique of its present-day situation, which the author also uses structurally in the context of *Die Rättin*. Once again in Garde's words:

> And so in *Die Rättin*, as prophesied in the other truth of *Der Butt* fairy-tale . . . the male's instinctive drive not only to dominate nature, but to reshape it himself, results in catastrophe. In that case Grass chooses

nuclear overkill for the end of the world, rather than a natural disaster as suggested in *Der Butt*.[29]

Nevertheless it is the imminent disaster of nature that leads to the resumption of the flounder story in *Die Rättin*. Five women are on a research assignment to quantify statistically the amount of jellyfish in the Baltic, a task which must be carried out on an especially equipped ship. In their marine isolation they therefore constitute a female community, if a somewhat precarious one. Any ambiguity is resolved when the ship is symbolically named the *New Ilsebil*. There is no need to cite the mytho-magical function of naming to understand this as an act to establish a female identity that goes with emancipation and the will for a specific female power.[30]

It is ironic that, despite their feminist background, their scientific work differs in no way from the approach of their male colleagues. This undermines their implicit claim to another way right from the start.[31] So even their appearance as knitting *Moirae* who spin the fatal thread of humankind is only a resigned rejection of former illusions. Referring to the flounder, the helmswoman says:

> It was about here we caught the flounder in the early seventies . . . Too bloody cocksure! Lots of hope and delightful promises. Nothing came of it though . . . Nothing changed. The men are still in charge . . . And we thought it was going to happen, all that women stuff, the clever rule of women. No way! (*R*, 65)

Their will to get involved has disappeared with their hope of bringing about changes.[32] They are just as content with strategies that soothe their conscience as the criticized male counterpart: 'their anti-conscience was buried in files and the statistical listings were ready next to the wool baskets' (*R*, 38).

The final bolt-hole left to the women after the failure of emancipation is Vineta: 'According to legend Vineta was a powerful and rich town, which perished because of the decadence and godlessness of its inhabitants. Grass and his feminists reinterpret Vineta as a matriarchical culture that has declined.'[33] This 'Utopia Atlantis Vineta' (*R*, 100) and its promise of a female empire 'against male power' (*R*, 99) becomes the last escapist solution after the collapse of all the ideals that still dominated *Der Butt*. The women now give up all practical involvement in politics and society, and turn their back on reality once and for all. But even their last hope is to be destroyed. When the women reach Vineta, the submerged town is populated by rats. Even before they burn up in the fierce glare of the nuclear explosion they realize 'that they have no place on earth ... that there is no refuge anywhere' (*R*, 324). The narrative of the *New Ilsebil* destroys not only its inherent ideals, but, by referring back to and thereby modifying its hypotext *Der Butt*, also the history of the German women's movement since the 1970s. The destructive force of both the Oskar Matzerath narrative and that concerning the *New Ilsebil* reverses Benjamin's notion of remembrance in the cruellest, most distorting way imaginable. Instead of the present making the past meaningful and thereby saving it from oblivion, its meaning and existence are retrospectively erased.[34]

In contrast to the above-mentioned examples, fairy-tales as folklore refer to the oral tradition of story-telling. They usually have a fairly simple structure, as the narrative is based on a dialectic that is finally resolved in a harmonious ending.[35] That said, even the Grimms' fairy stories (1812–15) do have a political function: the retreat into the past is supposed to conjure up a German national utopia against French predominance. The Romantic fairy-tale characters and the overall motif of the German forest are thus closely linked to the question of German identity. No wonder, then, that the theme of the dying forests has implications not only for fairy-tales, but also for the representation of national identity. The chancellor's family is driving along scenery that hides the dying forests, and the representative of the state uses his authority in a symbolic act to declare this theatrical world of appearances to be real, even though it is under threat of imminent collapse. Since contact with the world of nature only takes place via the media, nobody except the chancellor's children seems to question that version of truth.[36] The allusive character[37] of this stage-managed experience of nature has the function of sedating potential insecurities by means of Romantic clichés. Only the chancellor's children are able to see through and criticize this second-hand reality. They decide to drop out of this world of blindness and lies: they 'pull faces at their father and mother and on top of that at the TV teams and go into the forest of their own free will, mocking the Grimm version' (*R*, 56). In this act of self-determination the fairy-tale structure is simultaneously cited and negated. In the resumption of the fairy-tale it is once again social conditions that make the children go their own way. But while, in the fairy-tale, poverty and the wicked stepmother become the catalyst for action, in this case it is uncritical and self-destructive consumerism. While, in the fairy-tale, the movement from society to the forest does indicate a movement from security to danger, it is now the behaviour of society that causes the dangers, whereas nature appears to be the innocent victim of human possessiveness. In this way the resumed fairy-tale activates a critique of society. But whether it can really function as a corrective to society remains doubtful. On the one hand the authors, the Grimm brothers, are prestigious members of Parliament, but on the other they are much too unworldly to be properly influential in any political decisions.[38] It thus becomes the task of the fairy-tale characters to revolt against the imminent catastrophe.[39] Their walk to Bonn is based on their belief in the power of the persuasive argument, but fails because of the way things are. 'Unfortunately we have no power', says Jakob Grimm. 'Democracy is only a petitioner. Big money is where the power is!' (R, 236). The only thing left is 'the rebellion of the fairy-tale characters against the people'.[40] Once more the story can prove its power over people by magically sending the government off to sleep in order to protect nature and humankind against people. The 'anti-power' (R, 368) of the 'fairy-tale government' (R, 369) seems to be able to stop the catastrophe from happening, until suddenly Prince Charming, whose only right to existence lies in his kissing the Princess, mechanically starts his job and reawakens the government and its catastrophic tendencies. This blind actionism that shows neither brain nor conscience, and has no idea what consequences the actions may have, is not only a modern social development but already a possible narrative within the fairy-tale tradition. The moment this narrative can take over, catastrophe becomes inevitable. The final destruction of the fairy-tales is its result.

Now the narrative wheel has come full circle. In the face of catastrophe there are only 'stories that search for their ending' (R, 336), and even the new, post-human fairy-tales, 'in which the old ones miraculously survive' (R, 341), can only maintain the power of story-telling in a despairing fashion that contradicts all the probabilities. The narrative method which I have attempted to delineate does seem to have structural similarities with postmodern narrative practices. In stark contrast to the conventions of so-called postmodernism, however, far from welcoming this post-final condition, Grass perceives it as a cruel consequence of the violent end of human existence. His end of history refers to the death of humanity and has little in common with the reckless playfulness of a *posthistoire*. Even as he describes indifference in the face of catastrophe, he still wants to pursue and retrieve any remains of hope in which might reside some possibility of change. Grass insists not only on the importance of tradition and its critical incorporation into discourse, but also on the necessity of a concept of reason which undertakes a continual process of self-enlighten-ment.[41] Despite the apparent disappearance of the subject and the constructive function of its history, there is a powerful reluctance to recognize this development as necessary and inevitable within the text. For Grass, writing itself holds an inexhaustible remnant of hope.[42]

Notes

[1.] Günter Grass, *Die Rättin* [*The Rat*] (Darmstadt and Neuwied, 1986). Subsequent references will be indicated by the abbreviation *R*. Although this novel has been translated into English as *The Rat*, tr. Ralph Manheim (San Diego, New York and London, 1987), the original German text has been used for this detailed analysis. Translations of this and all other German sources are mine.

[2.] There are five narratives that sometimes intermingle, but can generally be kept apart. The sixth category is constituted by the poems that run through the book: the dream narrative of the rat and the anti-narrative of the narrator/author (1), the narrative that focuses on Oskar Matzerath (2), the narrative that focuses on the *New Ilsebil* (3), the fairy-tale narrative (4), and finally the Malskat narrative (5). (Thomas W. Kniesche, *Die Genealogie der Post-Apokalypse: Günter Grass' Die Rättin* (Vienna, 1991), 18)

[3.] Symbol is here understood in the sense of Cassirer. See Ernst Cassirer, *Wesen und Wirkung des Symbolbegriffs*, 8th edn. (Darmstadt, 1994), 79.

[4.] Kniesche, *Genealogie*, 208ff.

[5.] In Kniesche's view, the category of the apocalypse is unsuitable to describe the book and should be replaced by the term 'past-doomsday-story'. See Kniesche, *Genealogie*, 28f., 205f.

[6.] Günter Grass, *Werkausgabe in zehn Bänden,* ed. Volker Neuhaus, 10, *Gespräche mit Günter Grass* (Darmstadt and Neuwied, 1987), 348.

[7.] Grass, *Werkausgabe*, 10, 367f.

[8.] Grass quotes the title of a work by Lessing several times, which – as an argument with some of Reimarus's opinions concerning the Old Testament – tries to reconcile the oppositional eighteenth-century concepts of revelation and rationality. See Gotthold Ephraim Lessing, *Werke*, ed. Kurt Wölfel, *Schriften II* (Frankfurt am Main, 1986), 544–63.

[9.] In contrast to that is the picture of an ideal chancellor that the narrator invents. See *R*, 92.

[10.] Kniesche, *Genealogie*, 23.

[11.] See *R*, 167 and Grass, *Werkausgabe*, 10, 361.

[12.] *R*, 72–8 (description of the nuclear catastrophe), 85ff. (warnings that are ignored), 123f. (saying goodbye), 169 (end-game), 203f. (end), 204ff. (description of the end by the rat), 330 (end of history), 340 (end of all stories), 470 (disappearance of the human species).

[13.] Grass, *Werkausgabe*, 10, 346.

[14.] Genesis, 1:28.

[15.] Grass, *Werkausgabe*, 10, 365, and notes on 432f.

[16.] See ibid., 343, 346.

[17] See *R*, 141. This might be indicative of the affinity with Camus which Grass mentions in the Pinkerneil interview. See Grass, *Werkausgabe*, 10, 350.

[18.] In this respect the narrator finds himself in a position similar to that of Scheherazade in *The One Thousand and One Nights*.

[19.] Grass, *Werkausgabe*, 10, 350.

[20.] Ibid., 365f.

[21.] Gérard Genette, *Palimpseste. Die Literatur auf zweiter Stufe* (Frankfurt am Main, 1993).

[22.] Ibid., 18.

[23.] See ibid., 278 and 282.

[24.] Ibid., 278.

[25.] Ibid., 403.

[26.] Dieter Stolz, *Vom privaten Motivkomplex zum poetischen Weltentwurf: Konstanten und Entwicklungen im literarischen Werk von Günter Grass* (Würzburg, 1994), 316.

[27.] For this section I am indebted to Barbara Garde, *Selbst wenn die Welt unterginge, würden deine Weibergeschichten nicht aufhören. Zwischen Butt und Rättin: Frauen und Frauenbewegung bei Günter Grass* (Frankfurt am Main, Berne, New York and Paris, 1988).

[28.] Ibid., 9.

[29.] Ibid., 89.

[30.] For the function of mythological and magic rituals see ibid., 91f.

[31.] *R*, 22f.; Garde, *Selbst wenn die Welt* . . ., 94.

[32.] Stolz, *Vom privaten Motivkomplex* . . ., 338.

[33.] Garde, *Selbst wenn die Welt* . . ., 263.

[34.] See Walter Benjamin, *Gesammelte Schriften*, ed. Rolf Tiedemann und

Herrmann Schweppenhäuser, V, 1 (Frankfurt am Main, 1972), 588ff.; and Ernst Bloch, *Thomas Müntzer als Theologe der Revolution* (Frankfurt am Main, 1985), 14.

[35.] For differences between fairy-tales and myth, see Franz Fühmann, *Marsyas. Mythos und Traum* (Leipzig, 1993), 408–15, 446–8.

[36.] Grass, *Werkausgabe*, 10, 355.

[37.] See *R*, 54 and Kniesche, *Genealogie*, 13.

[38.] See *R*, 120ff. and Garde, *Selbst wenn die Welt* . . ., 258.

[39.] The fairy-tale community, under the leadership of the witch, represents a matriarchal structure that is in total contrast to the established forms of rule. See: *R*, 125–32 and Garde, *Selbst wenn die Welt* . . ., 255f.

[40.] Grass, *Werkausgabe*, 10, 356.

[41.] For parallels to critcs of postmodernism, see Burghardt Schmidt, *Postmoderne – Strategien des Vergessens* (Darmstadt and Neuwied, 1986). As a contrast, see 'Die wahren Feinde des Romans sind Plot, Charakter, Schauplatz und Thema: John Hawkes im Gespräch mit John J. Enck', in Utz Riese (ed.) *Falsche Dokumente. Postmoderne Texte aus den USA* (Leipzig, 1993), 331–45.

[42.] Grass, *Werkausgabe*, 10, 350.

11 • The Green *Bildungsroman*

Bill Niven

The term *Bildungsroman* is traditionally used to apply to novels in which characters learn to adapt to the norms of society. In this chapter I suggest that modern writers in the German-speaking world frequently focus on the portrayal of the opposite process: namely the gradual disengagement of a fully adapted individual from the dominant social, intellectual and professional norms. This process is one of *unlearning*, of coming to question and even discard accepted wisdoms. To underpin the above thesis I will concentrate on three recent works which I believe exemplify best this inversion of the traditional pattern of the *Bildungsroman*: Max Frisch's 'report' *Homo faber* (1957),[1] Uwe Timm's novel *The Snake-Tree* (1986)[2] and Friedrich Cramer's novella *Amazonas* (1991).[3] The implication in these works is that the modern western values in accordance with which the protagonists have been leading their lives are of questionable worth: while for nineteenth-century writers such as Stifter and Keller, acceptance of the established social and moral code was a process worthy of emulation, Frisch, Timm and Cramer prefer to extract their protagonists from the matrix of normality and, in so doing, reveal to them and the reader the narrow, confining, even destructive nature of this matrix. These three works, however, do not merely, as it were, unravel the fabric of integration which the traditional *Bildungsroman* strives to create, but also orientate the protagonists towards a new set of values, values which I perceive to be essentially ecological. In this sense they are themselves *Bildungsromane*. And they strongly imply that such values are to be acquired not in civilized Europe or North America, but only in contact with the more 'primitive' peoples and jungle environment of countries in Central America (Guatemala in the case of *Homo faber*) and South America (Brazil in the case of *Amazonas*, Argentina in the case of *The Snake-Tree*).

The above works exhibit a considerable degree of ecological

awareness. Green philosophy arose from concern for the environment, and indeed concern at the destruction of the South American jungle plays a role in *The Snake-Tree* and above all in *Amazonas*, where F. tries to raise the level of ecological consciousness both in South America and back home in Germany. The left-wing thrust of green philosophy is motivated by opposition to the might of industry and capitalism, not just at home but in the Third World, where indigenous cultures are being exploited and gradually eroded by the impact of reckless industrialization programmes very often initiated and led by western firms and cartels. In *The Snake-Tree*, the paper-making factory in the South American jungle is built by a German firm, not just at the cost of trees, but also by exploiting cheap Indian labour. In both *Amazonas* and *The Snake-Tree*, Europeans and Americans live in post-colonial luxury in areas where natives are excluded. Where they are needed it is for labour and prostitution. The human as well as the natural environment of the Third World is endangered by western intrusion. The most important element of green thought in the above works, however, is scepticism towards the dominant western *Weltanschauung*, which, in its obsession with the technological control of nature, has come to see man as all-powerful master of his destiny. All three works demonstrate that man remains mortal and fallible for all his powers; indeed the very unrestricted nature of his might, far from rendering him supreme, leads to the danger of global destruction. And in striving for technical control of the world, man turns himself into a machine: functional efficiency becomes his personal as well as his professional goal. This efficiency can only be maintained by an essentially rational mode of existence in which psychological and emotional needs have little scope for expression and are generally suppressed. The result can be fatal, as in *Homo faber*, where Faber's stomach-cancer can be interpreted as a physical symptom of spiritual decay. It is therefore not only man's external environment which is under threat, but his own *internal* one.

The main protagonists of *Homo faber*, *The Snake-Tree* and *Amazonas*, Faber, Wagner and F. respectively, are at the outset successful, career-orientated men involved, either as engineers (Faber, Wagner) or at managerial level (F.), in large-scale technological projects. Each has little time for or interest in matters of the heart or spirit, subordinating these to the rational exigencies of the job. In the course of the respective works, the three characters undergo an at least partial transformation. They learn to doubt, to varying degrees, the scientific world-view with which they previously identified, recognizing or at least considering the existence of forces such as fate, the irrational, metaphysical, intuitive and unpredictable. The western idea that everything can or *should* be measured and controlled is no longer entirely acceptable to them. All undergo an emotional regeneration, discovering instinct, spontaneity and sensual awareness; and they ultimately come to abandon their lives of professional commitment in order to take more time for themselves.

The above works also suggest that, while the First World still tends to impose its values on the Third, the time may have come for a reversal of the trend, with the First learning from the Third rather than vice versa. It is worth stressing that Frisch, Timm and Cramer give a modern slant to a long-standing tradition of anti-colonization in German literature. These authors condemn the destruction wrought by colonization in the interests of technology and industry, not Christianity, but they link the technologists with the missionaries of the past. Marcel in *Homo faber* talks of 'the engineer as the last edition of the white missionary' (*HF*, 50); similarly, Juan in *The Snake-Tree* describes western engineers as 'missionaries' (*ST*, 140). These works can thus be seen as continuation of a tradition of anti-colonization which goes back as far as Heinrich Heine. In the poem *Vitzliputzli* (1846), Heine describes the Spanish invasion of Mexico. While dwelling on the bloody ritual sacrifices of the Aztecs, Heine also condemns the Spaniards, who act in bad faith, and Christianity and western civilization in general. In the twentieth century, Alfred Döblin in his trilogy *Amazonas* describes the destruction the Europeans bring to the Inca and other Indian peoples. In the second part of the trilogy, *The Blue Tiger* (1938), the missionary Las Casas has more hope of his Christian message being understood by the natives than by his fellow Spaniards, who murder mercilessly. Uwe Timm, in his novel *Morenga* (1978), shows the bloody impact of German colonization on South-West Africa. And Hans Christoph Buch, in his novel *The Wedding in Port au Prince* (1984), describes the German and French colonization of Haiti – the enslavement of the natives, the arrogance of the whites, and their destruction of everything they set their hands on. These works are concerned principally with the havoc caused by colonization, but the underlying implication is that the culture and values of the oppressed peoples are arguably of greater worth than those of the Europeans.

I would now like to consider Frisch's *Homo faber*. Faber is a UNESCO engineer responsible for technical projects in South America: he provides 'technical help for underdeveloped countries' (*HF*, 10). The basic implication of Faber's job is that the representatives of technical mastery function as efficiently as their technology. But during his journeys Faber is confronted both by the unreliability of technology (his plane crashes in the desert) and by his own helplessness when deprived of its support. Indeed, his continued dependency upon it renders his behaviour awkward, obsessive and absurd. Witness his desperate need to shave in the desert, where there is no electricity, or his insistence on cleaning Herbert's car during his second Guatemala visit: it is subsequently useless because surrounded by mud and elevated upon a raised platform – rather like a machine idol to be worshipped for its own sake. Faber has lost sight of technology's practical applicability or relevance.

Faber means 'maker', and the central ironic thrust of *Homo faber* is that Faber, in his blind professional commitment to rationality and technology, has become so obsessively one-dimensional and inflexible that he rides roughshod over his own life, ignoring his physical, sensual, emotional and spiritual needs. He thus, effectively, 'unmakes' himself in the same measure as he helps to build the technical world around him. Not only does Faber live as if in contempt of his own nature, contracting cancer as a result, but he is committed to a life-denying philosophy. He is convinced of the desirability of abolishing death by means of technology: 'primitive peoples tried to negate death by depicting the human body – our method is to replace the body. Technology instead of mysticism' (*HF*, 77). And he believes in preventing life: 'abortion: a consequence of culture' (*HF*, 106). In the course of his journeys, Faber undergoes a revitalization, culminating in his Dionysian celebration of life (*HF*, 181). The emotional impoverishment of his existence hitherto becomes clear to him, and he comes to realize not only that he cannot overcome the cycle of birth, life and death, but also that he does not want to: that, in other words, mortality is a precondition of experience (*HF*, 199). What causes Faber to develop as he does?

All relevant knowledge, so contemporary wisdom would have it, flows from the First to the Third World. It is the First, after all, that has declared the technical control of nature to be a *sine qua non*, and is bent on convincing the technically less advanced of their destitution. In *Homo faber*, Faber's first trip to the Guatemalan jungle demonstrates, however, that he has more to learn from a 'primitive' environment than he has to offer it. The jungle represents untamed nature: the cycle of life, death and rebirth is revealed in a most explicit way, as is the intimate connection between these. Faber is disgusted by 'this constant reproduction everywhere, it stinks of fertility, of blossoming decay' (*HF*, 51). But, like it or not, he is forced to confront the palpable primacy of the *female-driven* natural cycle – which the western and predominantly *masculine* principle of technical control strives to annul. Faber cannot but sense the significance of the jungle's message, and even describes himself and his companion Herbert as emerging from it 'slimy like new-born children' (*HF*, 69).

It can be argued that Faber learns not only from the jungle, but also from the indigenous Indians. Faber finds them too peaceful, applying the adjectives 'childish' and 'effeminate' to them: they are 'as motionless as mushrooms, satisfied without light' (*HF*, 38). But the jungle climate soon forces Faber to stop either thinking or moving, and while he finds this experience unpleasant, it represents a kind of liberation from his machine-like existence. Much later, when in Greece, Faber is able to appreciate fully the pleasure of inactivity, he and Sabeth spending all night on the beach, looking up at the stars, themselves quite happy without electric light. It is from the Indios that Faber perhaps first intuits the joy of life as it is, beyond all technology, modern civilization and progress, which is central to the Sabeth episodes. And in rediscovering emotion, playfulness and curiosity with Sabeth, Faber comes to resemble the Indians by activating the woman and the child inside him so that he is no longer just 'man'. Similarly, F. in Cramer's *Amazonas* comes to see himself, in a series of mirror-images, as a combination of man, woman and child (*AM*, 59f.). Towards the end of *Homo faber*, Faber travels to Cuba, and his adulation of the Cubans contrasts with his attack on the artificiality and superficiality of the American way of life. In contrast to the pale white Americans, the Cubans have retained something natural, instinctive, animal-like, and these qualities are directly related by Faber to what he perceives to be their joy of life. It is, then – in complete contrast to western notions of the superiority of white over black – from *dark-skinned* peoples that western man can learn.

Most analyses of *Homo faber* rightly point to Faber's relationship with Sabeth as the experience which transforms him, stimulating his emotional and sensual nature. But the Guatemalan jungle prepares the way for this transformation. Frisch stresses this by placing Faber's description of his first meeting with Sabeth on board ship immediately after his account of the departure from Guatemala. The connection between these episodes is underpinned by Faber's response to the on-board party. The dancers' movements seem to him overtly sexual: repulsed, he compares the sultry atmosphere to that of the jungle (*HF*, 89), denounces feeling and disclaims any interest in marriage. Yet he already loves Sabeth, and before he leaves ship he proposes to her. Faber is thus drawn into an intimacy anticipated by the atmosphere of the Guatemalan jungle. It is also true that it is through the relationship with Sabeth, who turns out to be his daughter, that Faber is forced to confront the supremacy of the natural cycle: Sabeth was born despite his wish that she be aborted, his love for her more or less overcomes all rational and moral scruples, and she dies at the height of their love, his own death following shortly afterwards. But it was the jungle that first introduced him to this cycle: and it is the jungle which he subsequently associates with it (e.g. 'only the jungle breeds and decays as nature wants', *HF*, 106); and Marcel actually links the jungle, its atmosphere of death and reproduction, with woman (*HF*, 69), positing a parallel between them.

Like Faber, Wagner in Timm's *The Snake-Tree* is a typical representative of the technical generation. A successful engineer, he comes from Germany to South America with the task of rescuing a construction project which is threatened by corruption, self-interest, local politics and slothfulness. Initially determined to succeed against all the odds, he tries to change the way things are run wherever he can. The Bolivian workers call him 'big foot' (*ST*, 141), while another German involved in the project, Steinhorst, tells him to stop interfering (*ST*, 74). But when he ultimately does abandon his attempt to complete the project, it is not because of the pressure put on him, nor because of a corrupt interest in its non-completion, but because he has come to believe in the need for letting things develop as they will rather than forcing order and control upon them. That this decision is a point of principle is made clear at the end when

Wagner refuses money as a reward for his decision, and distances himself from the other Europeans involved in the project by referring to himself as someone who 'acted out of conviction' (*ST*, 311). Just as he comes to spurn control in his job, so he releases it in his emotional life. He has an invigorating affair with his young Spanish teacher, and, after some prevarication, sends a letter to his wife suggesting separation after years of unhappy marriage. Timm draws parallels between the technical domination of external nature and the repression of the self so typical of western society. Wagner's role in the construction of the paper factory was to shore up a collapsing project by quite literally stabilizing shaky foundations. For years, too, he had been shoring up his marriage. By the end of the novel his concept of progress has changed. Now it can only be achieved by allowing things the freedom to collapse. Wagner has no idea as to the direction his life should now take: it is his hope that, in releasing control, events will point the way forward (*ST*, 302). That collapse and chaos can be potentially creative is mirrored by the natural and social upheavals which accompany and finally merge with Wagner's development. The apocalyptic flood at the end of the novel suggests a *tabula rasa*, a wiping away as a prerequisite for rebirth and regeneration. In approaching the green hill, residence of generations of European colonists and emigrants, the flood promises to end centuries of intrusion that have brought destruction and exploitation to South America: the green hill has flourished, while the surrounding jungle has been transformed into red desert as the result of deforestation. At the same time as nature reclaims its territory, the local people are rebelling against the military junta. Both nature and the indigenous population are demanding the right to self-determination. As if to pour scorn on western civilization, fat black cockroaches emerge from the taps and plop into Wagner's bathtub – the jungle is returning.

Wagner's development in the novel is conceived of in terms of the contrast between the colours yellow and green. He initially identifies with the red and yellow moon landscape left behind by the technological destruction of the jungle; forests, he claims, are boring (*ST*, 138). His firm, which advertises its contribution to progress on television by showing a bulldozer destroying forest to the accompaniment of Beethoven's setting of Schiller's 'Ode to Joy', is represented by a red and yellow symbol. Elsewhere in the novel, yellow is associated with plastic helmets, boots and tablecloths. It is the colour of the man-made, synthetic and unnatural; and it is associated with intrusion. When Wagner angers the Bolivian workers by shoving a hand into their food, he does so because he is fascinated by the yellow powder they use as seasoning. He disregards what the natives regard as boundaries of propriety. But it is his disregard for something *green* which is to prove fatal. At the beginning of the novel, he runs over an emerald-green snake rather than risk damaging his car by swerving: he puts technology before nature. The snake is sacred to the natives, however, who worship animals and birds. According to the natives, anyone who kills this snake will die of drowning. From almost the first pages of the book, Wagner's fate is sealed: although he does not actually die, nature is to have its revenge on the brash indifference of the representatives of technology to a nature-based order. That the killing of the snake triggers an anti-technology dynamic is obvious from the inscribing of a snake symbol on Wagner's car: it is a tattoo of vengeance directed as much at the vehicle as at Wagner. But by the end of the novel, Wagner himself has distanced himself from the western principles he once espoused. He expresses the wish to put the brakes on the train of progress he has been responsible for driving (*ST*, 296). Increasingly, he is seen in terms of the colour 'green'.

Arriving in South America and realizing that the sun sets and rises at a different point, Wagner recognizes that he will have to change his way of thinking (*ST*, 14). Then, driving through the jungle, he wants to get out of the car and feel the enveloping greenery around him. This first intrusion of nature into his thoughts is underpinned by Timm's description of the roots of the trees running beneath the asphalt of the road and, at points, breaking through it (*ST*, 29): just as nature gradually breaks free from the superimposed road, symbol of linear thinking and technical control, so it is making inroads into Wagner's consciousness. Shortly after this he runs over the snake. The snake comes to symbolize Wagner's trapped position in life. He later recalls the incident, seeing the snake in his mind's eye, its body crushed but still struggling forwards towards the protective green of the trees (*ST*, 42). The dense jungle stands for natural growth, for the free and unimpeded. Towards the end of the novel, Wagner returns from a business trip to the capital. Forced to leave his car and traverse the jungle on a donkey, Wagner, stripped of technology, is directly exposed to nature and marvels at the manifold tones of green (*ST*, 272). For the first time he sees technology in a context which renders it apparently senseless. Thus the stolen distributor of his car is returned to him long after he has abandoned the vehicle, so that the distributor is effectively useless. He chances upon a motorway intersection without a motorway, and crosses a multi-lane bridge that has no connecting roads and is half overgrown with greenery. These uncompleted technical projects anticipate and in part motivate Wagner's abandonment of the paper-factory project. While reasons for non-completion of such projects in South America have more to do with political and economic instability than with the power of the jungle, the *impression* is of nature breaking the hold of technology and reclaiming lost territory. Nature's self-assertion accompanies and conditions the reassertion in Wagner of his own nature – spontaneity, sensuality – against the exercise of control in his professional and emotional life. It is during this trip back from the waterfalls that he decides to end his marriage and to abandon the paper-factory project. Nature is to hold sway.

I should like finally to consider the most recent of the three works, Friedrich Cramer's novella *Amazonas* (1991). The novella attempts to reconstruct and understand, in the course of a conversation between his friend Detmar and lover Eva, the fate of a certain F. A successful businessman, he is about to be appointed to the board of directors of his firm when he abandons his managerial career and goes to Brazil to work as an environmental technologist. Hoping to find varieties of fish that can be easily bred in the cooling basins of a power station, F. undertakes a long journey up the Amazon. It is also a journey to the end of civilization, a journey into the depths of the jungle and of the culture and peoples indigenous to it. The novella ends with F.'s death as a result of a snake-bite. This snake is named after one of the Fates, *Lachesis mutus*. Ironically, this is the Fate responsible for spinning the thread of life, not for severing it (*AM*, 86). The irony is intentional. F's death is, on a higher level, a liberating experience, the most extreme expression of life. He writes on a postcard to Eva: 'death is the sum of all feelings' (*AM*, 47). As in *The Snake-Tree*, death constitutes the ultimate escape from the emotional restraints of the civilized western life-style.

According to Detmar, it is from the Indians that F. learned the importance of feeling: 'the spontaneity of these people overwhelmed him' (*AM*, 46). While western civilization is enlightened, intelligent, cynical, Detmar continues, it has lost the ability to register and read emotion: 'in the realm of feeling we have become illiterate' (ibid.). In Cramer's novella, it is the white Europeans who appear primitive in comparison with the Indians, to whom this term is conventionally applied. The Indians are not only in contact with their feelings, they also retain a closeness to nature, manifest in an empathic understanding of their environment and in an ability to intuit coming events. In the novella, developments which cannot be predicted by western thinking, which is based on causal and rational networks, are sensed by the Indians, whose awareness depends on instinct and experience. In a memorable scene, the narrator describes how western scientists, despite using sophisticated sonar equipment, cannot locate any shoals of fish. The natives laugh at their attempts, and then take them to where the fish really are (*AM*, 92-3). The implication is that technology, no matter how fine, can never replace indigenous experience. Modern technical culture fails in another sense to match up to that of more 'primitive' cultures: it severs the nexus between man and nature that is the only guarantor of stable ecology and world survival. When the natives in Cramer's *Amazonas* burn down forest to create clearings, they do so according to a system that prevents erosion. They lay bare only a small area and move on after two years before all the nutrient has been leached out of the soil. The forest can then grow back (*AM*, 86-7). What a contrast to the industrial slaughter of square mile upon square mile of jungle, described in Cramer's novella in mind-boggling images of huge tractors, a massive chain spanned between them, marauding their way through the jungle and destroying any chance of nature's regeneration.

The jungle and its inhabitants make a new man of F., revitalizing his instincts and creating a sense of solidarity with the natives and with both organic and inorganic nature (*AM*, 114). But it is too late for him to start his life over again, as it is too late for Faber and Wagner. Indeed, both Eva and Detmar suggest on different occasions (*AM*, 22 and 114) that there is an intimate connection between F.'s revitalization and his subsequent death, as if such a renewal is only possible at the cost of life. All three works discussed here seem to posit the impossibility of integrating this renewal into life – rather as if the protagonists had entered a sphere of experience too intense for their emotionally impoverished western constitutions to withstand for long. One could therefore argue that the works end pessimistically. But *Homo faber* and *The Snake-Tree* leave the actual moment of death unstated, a formal open-endedness transcending the mortality of the heroes. And death is not necessarily negative: for F. it is the sum of all feeling, while in *The Snake-Tree* it is associated with an apocalyptic sweeping away of corrupt Europeanism and the possibility of natural and social revival. Moreover, the mortality of the protagonists is evidence of their fallibility, of the ultimate powerlessness of technology; it is punishment for their male hubris, a nemesis set in motion by fate (*Homo faber*) or Indian superstition (*The Snake-Tree*). Eva in *Amazonas* even goes so far as to conceive of F.'s death in terms of a 'sacrifice' or 'penance for . . . our civilization' (*AM*, 39). The conclusion of the works suggests the existence of a metaphysical force which might restore the balance upset by technology, a message, given the destructiveness of that technology, of hope.

One might also ask the question as to what extent the protagonists really develop. Faber's experiences in Mexico and Guatemala and his love affair with Sabeth are described in a series of flashbacks in which Faber continues to assert his philosophy of man's mastery over nature; Wagner's policy of non-interference could be construed as passivity; and F.'s support of environmental issues seems to be contradicted by the fact that he offers advice to industry in South America. But a Walter Faber who rejected overnight the principles on which his life was based would not be credible. *Homo faber* is a psychologically realistic portrait. Change is experienced in terms of the dichotomy between old and new, rational and emotional, a dichotomy also captured in the increasing tension between text and subtext. As for Wagner, his non-interference has to be seen in context: doing nothing is an act of rebellion precisely because western values are based on *doing*. Doing has become an end in itself, divorced from issues of inner necessity: *non-doing* must be the first stage in any process towards the rediscovery of will-centred doing. In the case of F., the apparent contradiction is resolved by F's realization, hinted at by Eva (*AM*, 20), that ecological progress can only be achieved together with industry, not in opposition to it.

The French botanist Aimé Bonpland, who accompanied Alexander von Humboldt on his trip to South America, underwent an extraordinary transition. After looking after the Empress

Josephine's gardens at Malmaison outside Paris, Bonpland returned to South America. There he lived in a simple hut of pampas grass, looked after by a native by whom he had several children. He slept on the ground and eventually no longer knew how to use a knife or a fork. In Friedrich Cramer's *Amazonas*, we meet a similar character, Heidi, who has withdrawn in South America from all contact with the civilized world. Her only friends are sloths. In the course of time she has forgotten how to speak. And in *Homo faber*, Faber's travelling companion Herbert remains in Guatemala, permanently cut off from civilization, and develops a temperament of grinning indolence. Neither Heidi nor Herbert are figures either Cramer or Frisch – or Timm – would hold up to us as examples to follow. Their works are not about becoming like the natives, but about learning from them and their jungle environment, about taking some of their understanding of life over into western thinking. Nor are their works about dispensing with technology. Thus Eva, in Cramer's *Amazonas*, talks rather of a post-technical era, where technology is no longer an end in itself but really there to serve mankind (*AM*, 79). She calls for a new asceticism, an attitude to technology shaped by a genuine awareness of its limits and destructive potential (*AM*, 79). The works discussed in this chapter are fundamentally ecological not just because they express indignation at the destruction of the green lung of the earth, but because they advocate a return to living together with both internal *spiritual* and external *physical* nature. It is this emotional and material ecology which is central to the lives of the natives.

Notes

[1] All references to *Homo faber* are taken from Max Frisch, *Homo faber* (Frankfurt am Main, 1977). Translations are mine and references are indicated by the abbreviation *HF*.

[2] All references to *The Snake-Tree* are taken from Uwe Timm, *Der Schlangenbaum* (Cologne, 1989). Translations are mine and references are indicated by the abbreviation *ST*.

[3] All references to *Amazonas* are taken from Friedrich Cramer, *Amazonas* (Frankfurt am Main, 1991). Translations are mine and references are indicated by the abbreviation *AM*.

12 • Literature and the Environment in the GDR: Some Implications of Post-1989 Disclosures

Martin Kane

In their inquiry into the cost of German unification, Jan Priewe and Rudolf Hickel noted that the GDR could only sustain what was an already inefficient economy by over-exploiting and inflicting extreme damage on the environment and the country's natural resources.[1] What they leave out of their account of the ensuing catastrophic ecological legacy is that the shortcomings in the systems of industrial production in the GDR were also highly exploitative of the work-force and harmful to its health and well-being. This chapter concentrates on two GDR novels, Werner Bräunig's *Rummelplatz* [*Fairground*][2] and Martin Viertel's *Sankt Urban* [*Saint Urban*],[3] which deal with the lives of workers in the mining regions of Saxony and Thuringia – the *SDAG-Wismut*, the 'state within the state',[4] set up by the Soviet Union in the immediate post-war period to satisfy its desperate need for uranium. It will examine the extent to which the imperatives of the world of work, and how they were to be represented in literature, meant ignoring, or giving only oblique consideration to, the hazards of working in this industry.

Ecology issues were always a taboo, or quasi-taboo subject for GDR writers. They had to be tackled with circumspection and liberal use of the 'Schere im Kopf', the censoring scissors in the writer's own head. What might happen if a writer ignored this mechanism is illustrated by the case of Hanns Cibulka's *Swantow*. A first draft had been published in *Neue Deutsche Literatur* (April 1981), but the book version could only appear after considerable reworking of the text and the deleting of references to the effects on water, air and food of chemical and nuclear pollutants.[5] Needless to say, works which insisted on unforgiving and explicit images of environmental pollution in the GDR – the shoals of eels on the Baltic coast, stranded in seaweed and twitching their last, in Christa Moog's 'Die Saison 80 in Göhren begann' ['The 1980 season in Göhren began'],[6] for instance, or the baleful account of air heavy with toxic industrial filth in Monika Maron's novel *Flugasche* [*Flight of Ashes*] of 1981,[7] could only be published in the West.

None the less, and despite censorship, literature in the GDR was more forthcoming about environmental problems than were official sources of information.[8] From the early 1970s on, what begins as the articulation of doubts about the ability of technology and science to deliver the socialist utopia develops, in the work of Hanns Cibulka, Christa Wolf, Jurij Brězan, and others, into discussion about the increasingly irreconcilable claims of economic growth on the one hand, and sensible environmental policies on the other. Indeed, given the hindrances to their publication, the amount and variety of works on the increasingly parlous state of the environment is surprising. In 1989 Steffen Peltsch drew attention to some forty-two novels and stories, some 150 poems, and texts in eight anthologies which had been published since 1980 on ecological themes.[9] He also comments on an earlier flourishing of an ecological literature for children, reminding us that writers of fiction for children were often permitted greater licence in the treatment of ecological questions than those addressing an adult audience.[10]

In view of the range of perspectives on pollution in the GDR which it opens up so dramatically, it would be appropriate to preface discussion of the works by Bräunig and Viertel with some observations on what is arguably the GDR's most celebrated environmental novel, Monika Maron's *Flugasche*. The novel is both revelatory about the fact of industrial pollution in the GDR, and, through the combative stance of the first-person narrator, strong on indignation: 'This miserable hole ... B. is the dirtiest town in Europe.'[11] But it also – in the exchanges, real and imaginary, between the crusading journalist Josefa Nadler and her immediate boss and mentor Luise, as well as in her later meeting with a highly placed bureaucrat – presents us with the official case for the problem remaining strictly under wraps.

Despite her apparent initial endorsement of the piece which Josefa has written about B. (clearly Bitterfeld), Luise advises her to tone down her approach. Her arguments stem from a reluctance to diminish what she sees as having been achieved in the GDR; they are the result of measuring the 'advantages of socialism' against the barbarities of the past. The views of the party official, on the other hand, are rooted in an entirely pragmatic view of industrial and economic conditions in the GDR. Since nothing can be done to remedy the problem of pollution, the fact of it should not be voiced at all. In delivering a gentle but firm reproach to Josefa – aimed to bring her back on side – he formulates in restrained but unambiguous manner the philosophy which informed all official thinking in the GDR about pollution, industrial and nuclear.[12] In a state which constantly proclaimed itself in the slogan 'Alles zum Wohl des Menschen' [Everything for the good of the people], questions of economic and political import invariably took precedence over questions of individual welfare.

This point of view is explained by the industrial realities of the GDR. The state was committed from an early stage to economic growth based on the Soviet model of heavy industry, and, after the oil crisis of the 1970s, was forced to draw 70 per cent of its energy from lignite – in its production of high quantities of sulphur dioxide and airborne dust and ash one of the most polluting of all fuels. It is little wonder, therefore, that the GDR authorities were forced to turn a blind eye to the devastating environmental effects of pollution. It would have meant economic suicide to have done otherwise.

What is not treated directly in Maron's novel is that the establishment of this mode of industry was rooted in the belief in the early years of the GDR in the transforming and liberating effects of science and technology. This philosophy finds explicit and didactic endorsement in Karl-Heinz Jakobs's *Beschreibung eines Sommers* [*Description of a Summer*]. Here, in the camp-fire conversation of a group of enthusiastic young communists, a paean is delivered to the ability of science to move the earth – literally, as well as metaphorically. What barely a generation later would be revealed as man's fatal hubris and arrogance is here starry-eyed vision of a world in which the Gulf Stream can be redirected, the Sahara irrigated, and atomic science has the answer to all of mankind's problems. Nuclear technology, hand in hand with socialism, will transform the world.[13]

The conditions of work described by Maron, Josefa's experience of the sheer brutality of industrial labour, of course make a mockery of a view of the future proposed by Jakobs's novel in which atomic power and computer technology would remove all drudgery from the work process. Nevertheless, Luise's arguments have to be seen in the context of these earlier hopes and expectations. They do little, however, to persuade Josefa that she should withhold her article. In rehearsing the argument she is about to have with Luise, she formulates a deliberately provocative standpoint: 'Wem nutzen unsere Schwindeleien, Luise?' [Whom do our swindles help, Luise?] she asks.[14] The use of the word *Schwindelei* provides a clue to the ways in which the treatment in literature of the uneasy relation between landscape, nature and the incursions of industrialization were viewed by East Germany's writers and literary critics.

One seems here at times to be confronted with a conspiracy concocted for the purpose of propagandistic and ideological obfuscation. Franz Fabian, for instance, in 1975, offsets his account of the ravages perpetrated on man and nature by industrial capitalism with a vision of a socialist society in which nature is no longer the object of unbridled exploitation.[15] His observations are not original. He acknowledges a debt to Annemarie Auer and her much earlier description in *Landschaft der Dichter* [*Landscape of the Poets*] of 1958 of the writer's relationship to the ecological havoc wrought by industrialization, and the ways in which socialism might mitigate it: 'Not until our own times in the socialist countries does the urge arise, alongside the exploitation of natural resources, for an orderly shaping of the landscape.'[16]

How are such sanguine views of the relationship between socialism and industrial development to be regarded? Whether we take them on trust or prefer to be sceptical and interpret them as clouding of the obvious for ideological reasons is a matter of debate.

For at least one East German writer, however, Günter Kunert, there is no argument to be made for the superiority of socialist over capitalist modes of technological progress. In a frank exchange with Wilhelm Girnus in 1979 in *Sinn und Form,* Kunert dismisses the notion that in the hands of 'progressive forces' science and technology have none of the negative consequences they have under capitalism. That the effluent from industrial plant owned by the people is in some mysterious way less polluting of rivers and lakes than that which flows from capitalist enterprises is just as absurd as the notion that socialist car exhaust fumes may be preferable to capitalist ones.[17]

Where do these debates leave us with regard to literature depicting the mining of uranium? It might just about be possible to accept that, having committed the economy of the GDR in the late 1940s and early 1950s to the primacy of heavy industry, it was, by the 1970s and 1980s, too late to turn back, and that there was no alternative for the GDR but to try to ignore the devastating ecological consequences. It is difficult, however, in the light of what has emerged since the collapse of the GDR, to apply the same degree of forbearance to the mining of uranium and its repercussions. Articles in *Die Zeit*, the *Thüringer Allgemeine*, the supplement to *Das Parlament* and, not least, the *Observer*, demonstrate a catastrophic legacy. I detail it in the briefest of outline: forty years of pumping the radioactive waste product into a man-made lake larger than 300 football pitches and fifty metres deep bordering on the village of Oberrothenbach in Saxony has led to permanent pollution of the ground water and produces, as the lakes dry out in hot weather, a ubiquitous layer of radioactive dust. It has been calculated that some 22,500 tons of arsenic have been pumped into this lake, along with the uranium mud: 0.2 g is a lethal dose. Subjective registration of an abnormally high level of deaths from cancer in the district cannot be confirmed, because no records have been kept. Variations on this state of affairs repeat themselves in a triangular region stretching from Gera to Chemnitz in the east to Johanngeorgenstadt in the south. In all, forty-five years of mining which produced some 220,000 tons of uranium ore for the Soviet Union left 1,200 square kilometres of land permanently contaminated – an area equal to half of the Saarland.

How did this calamitous situation arise?[18] The Soviet Union had inadequate supplies of uranium of its own. In order to fuel its A-bomb it had to exploit sources in the mines in Saxony and Thuringia.[19] To this end, an area comprising thirteen districts and some 2.1 million inhabitants was brought under Soviet control at the beginning of 1946, and the highest priority was given to the reopening of largely disused uranium mines. The operation was detrimental, because of the cost, to other areas of the East German economy. It also inflicted considerable damage and disruption on the local landscape and population. Any reservations about the whole process, however, were silenced by subordinating them to the question of peace. Under the slogan 'Smash the American monopoly on the atomic bomb', the argument was made very simple. To preserve world peace, it was essential that the Soviet Union should have the bomb. Uranium was a vital prerequisite of this. Therefore it had to be mined, regardless of the cost, financial, environmental or human.

Both *Rummelplatz* and *Sankt Urban* reflect these imperatives and the urgency of the situation. They are given dramatic illustration in the second chapter of *Sankt Urban*, where Georg Bäumeling is woken from his bed by a Soviet officer and his sergeant and driven off to open up the closed mineshaft which gives the novel its name. And in the opening chapter of *Rummelplatz*, Herman Fischer recalls arriving at the mine, shortly after the war and at a moment of crucial historical significance for the preservation of world peace.[20]

The project was also subject to strict security. The setting up of areas of restricted access aggravated the problems caused by shortages of housing, labour, and particularly of technical expertise. The gaps in the work-force were filled by drafting in a rag-bag of ex-Nazis and soldiers, petty criminals and all those attracted by high wages and superior food rations. This all contributed to that atmosphere of at times barely regulated chaos and disorder which manifests itself in other areas of the rapidly expanding industrial economy of the GDR in the early years. It presents the East German writer, exhorted to dramatize these turbulent times as appealingly as possible, with certain difficulties: how, for instance, to reconcile brutal and potentially anarchic circumstances with a propagandistic presentation of the construction of socialism and the notion of the place of work as the setting for collective socialist endeavour? In other industrial novels such as Marchwitzka's *Roheisen* [*Pig Iron*] or Karl Mundstock's *Helle Nächte* [*Bright Nights*] these potential problems are glossed over with an element of socialist romantification. In *Roheisen* a character speaks of participating in the opening up of a 'New California'; in *Helle Nächte*, the narrator talks of having been transported to a place reminiscent of a 'Jack London gold-rush camp' (both references here are to the building of Eisenhüttenstadt, an industrial city carved out of agricultural and forest land near to the Polish border).

Marchwitza and Mundstock were returning communist exiles who wished to see the success at all costs of the socialist vision to which they had dedicated their lives and for which they had suffered. If, in literary terms, gilding of the lily was to be the price to be paid, then so be it. For Bräunig, who belonged to a later generation, things were not so simple: as both writer and political animal, he is something of a paradox. At points in his writing he is a model of socialist partisanship – he is, after all, the creator of the slogan under which the first Bitterfeld conference sailed – and fully able when he chooses to convey the ennobling effects of manual labour, that sense of the worker as hero, and of the place of work as where the individual may find complete social and political fulfilment. More frequently, however, the disparity between what his eye as a writer saw, and what his heart as a socialist wanted to see, presented him with serious problems. We see these conflicting elements scattered throughout his unfinished novel, but they are already visible in embryo in a poem based on his own experiences in the uranium mines and written while he was a student at the Johannes R. Becher Literary Institute in Leipzig:

> Zusammengepfercht in hölzernen Kisten
> auf Rädern –
> – Eisenbahn, Automobil –
> ziehen nach Süden farblose Gestalten
> in grauen Monturen,
> zerlumptem Zivil.
> Übriggebliebne, Entwurzelte, Leere,
> Ausgebrannte im Fieber der Nacht –
> ziehen nach Süden
>   ins Ungewisse –
> taufen verbissen den ersten Schacht.

[Herded together in wooden crates/ on wheels –/ – train, automobile –/ riding south, drab figures/ in grey overalls,/ ragged civvies./ Men stranded, rootless, empty,/ burnt out in the fever of the night –/ riding south/ into uncertainty –/ grimly baptizing the first shaft.][21]

The poem as a whole is a propagandistic celebration of uranium, but even in these opening lines we see pointers to those more sombre aspects which would be spelled out in the extract 'Rummelplatz' published in 1965 in *Neue Deutsche Literatur*. The existences adrift and rootless hinted at in these opening lines would be fleshed out in 'Rummelplatz' only for their author to fall victim to the repressive cultural climate of the SED Central Committee's eleventh plenary session of December 1965 and to be on the receiving end of Erich Honecker's excoriating comments about novels, Bräunig's amongst them, which were full of 'obscene detail' and gave a 'false, a distorted picture of the difficult beginnings for the uranium mining industry'.[22]

Unlike the 'heroic epic' created by Martin Viertel out of these events, which found official favour, the depiction in Bräunig's work of the social disruption caused by the setting up of the *SDAG-Wismut*, and the brutish and brutalizing circumstances in which those who worked in it lived was clearly all too graphic.

Owing in part to the furore it caused in 1965, in part to Manfred Behn's use of it in 1981 as the opening piece for his anthology of GDR literature, *Rummelplatz* is the more familiar of the two 'Wismut' novels under discussion here. Its story, along with that of its author, is one of the regrettable chapters of GDR literary history. Bräunig died prematurely at the age of forty-two in 1976, and as an alcoholic, his condition accelerated in all probability by the difficulties put in the way of what was his major work, which, had it been completed, promised to become one of the landmarks of GDR literature. Some 200 of the 368 sides of Bräunig's original manuscript were belatedly published in 1981, the selection being made, according to the editor, Heinz Sachs, on purely literary grounds.[23] Without access to that original manuscript it is impossible to know whether these were the only criteria. It is instructive, however, to compare the two scenes published in *Neue Deutsche Literatur* in 1965 and announced as being taken from an as yet uncompleted novel which would be published the following year with the versions of them which Sachs published. There are very noticeable differences. While the 1981 version still retains its sharpness in the portrayal of the uranium workers' lives, it has been pruned of several provocative passages. In addition to the 'state within a state' reference already mentioned, also removed are: an obscenity (see Honecker's remarks above); a critical comment on the physical appearance of the liberating Soviet Army; sceptical remarks about political opportunism and the attitudes of indifference of those in power to the lot of workers.

What we do not know of course is whether Bräunig himself, prompted by the criticisms made in December 1965, had made these deletions, or whether they are the work of Sachs. If the latter is the case, the question then arises whether Sachs had also deleted potentially controversial passages in Bräunig's original text dealing with the effects of uranium mining on both people and the environment. Any overly detailed references to neglect of safety precautions, to the effects of radiation, or the careless methods of waste disposal would clearly have caused difficulties in 1981, when these matters were increasingly a matter of debate, albeit semi-covertly.

In the novel as we have it, mention of the nuclear radiation which workers faced day by day at the rock face is negligible. There is an acknowledgement that elementary safety precautions are being neglected for reasons of speed and convenience. But never is there any real awareness of the true extent of the radiation hazards involved.

Is this to be explained by the 'scissors in the head'? Is it acceptance that the Soviet Union, for the sake of world peace, must have the bomb, and therefore an ample supply – at whatever cost – of the precious 'yellow cake', which keeps Bräunig's pen away from the topic of radiation? In part, perhaps, yes. But genuine ignorance must also be a factor. Even as late as 1992, and after a host of revelations, the philosophy in the uranium regions was still, according to the *Thüringer Allgemeine* article of 29 August 1992, to be summed up in the resigned comment that since one could not see, taste or smell radiation, then it could not be dangerous. If this was the case in 1992, how much more would it have been so in the 1950s when Bräunig began to write his novel?

What had shocked Honecker in the *Rummelplatz* extract was the depiction of uranium miners as spiritual desolates, without social or political purpose, seeking fleeting consolation in drink, sex and spurious excitement. The authenticity with which the utter bleakness of the fairground scene, and other episodes in the extended version are depicted is very important to this argument. It demonstrates Bräunig's inclination to go for reality in all its harshness, rather than to plough it for what might underpin his socialist convictions. The persuasiveness of these passages (and, incidentally, those that celebrate the act of physical toil) reinforces confidence in Bräunig as a realist. They lead to the conclusion that, as a writer, he was more devoted to the principle of realism than he was to the obligations of a partisan socialism. His failure to write about the true extent of the dangers of uranium mining, and to anticipate its horrendous long-term legacy, was not the product of ideological cowardice, but came about because he was not fully aware of them.

Martin Viertel, and *Sankt Urban*, have enjoyed none of the celebrity which has come Bräunig's way. The first that a wider audience in the West may have heard of him may well have been in the brief extract from the novel which was printed as part of a *Zeit*-Dossier in June 1991, one of the first disclosures of the ecological and human disaster left in the wake of more than four decades of uranium mining and the reckless disposal of the 90 per cent of it which was a highly radioactive waste-product.[24]

The irony of using an extract from Viertel's novel in this context was palpable. *Sankt Urban* had had a particular place in the literary mythology created out of the early years of the GDR; its subject matter, once viewed as 'The great theme: uranium mining, co-operation between Soviets and Germans after the war up until the founding of the GDR',[25] was now revealed to be, after Hiroshima and Chernobyl, at the heart of 'the third largest radiation disaster in history'.[26]

Sankt Urban is utterly free of the existential dimensions which make Bräunig's work so fascinating. While the scenery of his novel, the façades of his adventurer, 'gold-digger' uranium miners bear a superficial resemblance to those created by Bräunig, there the similarities end. This is in part, one suspects, due to the fact that Viertel is writing at a distance. Himself an ex-miner and graduate of the Johannes R. Becher Literary Institute, he is writing his novel in the late sixties with the clear intention of adding to the monument to the achievements of the early years of the GDR and its vanguard of communist activists. There are hints of the distinctly unsocialist chaos of these early beginnings, but the observation 'a handful of communists in a wild circus'[27] – uttered admittedly with a certain amount of scepticism – nevertheless reveals the novelist's ideological hand. As does his portrayal of the Soviets. In Bräunig's novel, the Soviets are there to give orders, to restore the peace in the wake of drunken brawls, they are the objects of suspicion. In Viertel, on the other hand, the presentation of Soviet impulse and guiding involvement is strictly orthodox. His hero, Georg Bäumeling, straight from the catalogue of worker activists with politically impeccable pasts, detects in the Soviet officer who has dragged him from his bed to inspect the disused mine shafts a sympathetic fellow spirit who inspires immediate trust and confidence.[28]

Environmental problems and the manifest damage being done to forest and landscape could clearly not be totally disregarded in any novel dealing with uranium mining. One of the impressive subtleties of Bräunig's novel is that not only are the ravages inflicted on the landscape given due expression, they are also made to reflect the inner landscapes of the characters. The desecrated physical landscape, the village threatened by subsidence, the local churchyard, cleared for a mineshaft never sunk, and then bulldozed into a square of flattened shale, are both setting and metaphor. They mirror the spiritual desolation of the bulk of his characters, who are washed up and without foothold in a society in turbulent transition, and quite unconsoled by slogans of the new dawn about to break.

In *Sankt Urban*, there is none of this. The reader is throughout fobbed off with ideological platitudes. On the one hand, the ecological issue, the damage to the forest by geological exploration, is justified in global terms: the Soviet need of uranium to preserve world peace, the struggle against world capitalism. What is a handful of trees in comparison to ideological victory over *IG-Farben*, *McCarthy* or *Clay* – just a few of the thousand names linked in devilish, capitalist purpose?[29] Or, attention is diverted from the ravaging of the countryside, by an anthropomorphization of the eponymous mine (we think inevitably of Zola's *Germinal*) in which the elevation of the struggle between miner and mine to at points mystical, pseudo-mythological proportions lets the reader forget what a baleful and insidiously hazardous process it actually was.

In evaluating and comparing these two works, one feels less magnanimous towards Viertel. Particularly in view of the fact that a new edition of his novel was published as late as 1985, when what the novel concealed of the damage done to environment and population was already beginning to emerge. In one of the didactic scenes in the novel, a young girl asks a mining engineer about the nature of the ore and the magic mountain from which it emanates. He replies: 'It gives off rays which cannot be seen or heard. But when they are controlled and put to use, man becomes a giant.'[30]

Clearly still subscribing here to the socialist myths about the benefits of technological progress, and still subordinating all to party purpose, would, one wonders, the author of *Sankt Urban* been so sanguine had he known about the child leukaemia victims of Ronneburg?[31]

Notes

1. Jan Priewe and Rudolf Hickel, *Der Preis der Einheit. Bilanz und Perspektiven der deutschen Vereinigung* (Frankfurt am Main, 1991), 43.
2. I refer to two versions of this text. The extract published in *Neue Deutsche Literatur*, 13, 10 (1965), 7–29, and reproduced in Manfred Behn (ed.), *Geschichten aus der Geschichte der DDR 1949–1979* (Darmstadt and

Neuwied, 1981), 7–22. The version published in Heinz Sachs (ed.), *Werner Bräunig. Ein Kranich am Himmel. Unbekanntes und Bekanntes* (Halle and Leipzig, 1981).

[3.] Martin Viertel, *Sankt Urban* (Berlin, 1968). Page references to the edition published by Verlag Tribüne, Berlin.

[4.] From the original version of 'Rummelplatz' in *Neue Deutsche Literatur* 1965. Notice that this expression is cut from the version edited by Heinz Sachs in *Werner Bräunig. Ein Kranich am Himmel*. Indeed the whole sentence, 'the uranium mines are a state within the state, and vodka is their national drink', has gone. This censorship of politically controversial material undermines the claim by Heinz Sachs that he has exercised his editorial function solely on the basis of aesthetic criteria.

[5.] For a full account of this see Ernest Wichner and Herbert Wiesner, *Zensur in der DDR. Geschichte, Praxis und 'Ästhetik' der Behinderung von Literatur* (Berlin, 1991), 107.

[6.] In Christa Moog, *Die Fans von Union* (Düsseldorf, 1985), 104.

[7.] Monika Maron, *Flugasche* (Frankfurt am Main, 1981).

[8.] Axel Goodbody is helpful here. See '"Es stirbt das Land an seinen Zwecken": writers, the environment and the green movement in the GDR', *German Life and Letters*, 47, 3 (1994), 325–36, here 326. Also his comment on 'the Council of Ministers' decision to keep environmental statistics secret in 1982' and on the 'failure of the GDR media to report the Chernobyl accident until well after the event', 334.

[9.] Steffen Peltsch, 'Schreibtischerwägungen', *Neue Deutsche Literatur*, 37, 11 (1989), 11–15, here 12–13.

[10.] See Hubertus Knabe, 'Zweifel an der Industriegesellschaft. Ökologische Kritik in der erzählenden DDR-Literatur', in Werner Gruhn et al., *Umweltprobleme und Umweltbewußtsein in der DDR* (Cologne, 1985), 201–50, here 214.

[11.] Maron, *Flugasche*, 16 and 36.

[12.] Ibid., 169–70.

[13.] Karl-Heinz Jakobs, *Beschreibung eines Sommers* (Berlin, 1963), 69ff.

[14.] Maron, *Flugasche*, 34.

[15.] In Franz Fabian (ed.), *Meine Landschaft. Prosa und Lyrik* (Berlin, 1975), 432.

[16.] Annemarie Auer, *Landschaft der Dichter* (Dresden, 1958), 57.

[17.] 'Anlässlich Ritsos. Ein Briefwechsel zwischen Günter Kunert und Wilhelm Girnus', *Sinn und Form*, 31, 4 (1979), 850–64, here 850.

[18.] A useful account with extensive bibliography is Rainer Karlsch, '"Ein Staat im Staate". Der Uranbergbau der Wismut AG in Sachsen und Thüringen', *Aus Politik und Zeitgeschichte. Beilage zur Wochenzeitung Das Parlament*, B 49-50/93, 3 December 1993, 14–23.

[19.] These areas had been occupied by American forces at the end of hostilities, but, bound by an earlier agreement, they had had to withdraw behind the Elbe at the beginning of July 1945. One wonders what would have become of the Soviet A-bomb effort had these areas of Thuringia and Saxony remained in American hands and not become part of the Soviet Occupied Zone.

[20.] Sachs (ed.), *Werner Bräunig*, 12.

[21.] Werner Bräunig, 'Uran – gewidmet den Kumpeln der SDAG Wismut zum V. Parteitag der Sozialistischen Einheitspartei Deutschlands', in *Ruf in den Tag. Jahrbuch des Instituts für Literatur Johannes R. Becher 1960* (Leipzig, 1960), 304–8, here 304.

[22.] In Elimar Schubbe, *Dokumente zur Kunst-, Literatur- und Kulturpolitik der SED* (Stuttgart, 1972), 1079.

[23.] Sachs (ed.), *Werner Bräunig*, 490.

[24.] Holger Douglas and Thomas Kleine-Brockhoff, 'Heiße Erde aus dem kalten Krieg', *Die Zeit*, 24, (7 June 1991), 15–17.

[25.] Eckart Krumbholz, 'Made in GDR: Martin Viertel', *Sonntag*, 36 (3 September 1983), 9.

[26.] See Thomas Rothbart, 'Die Wismut – der drittgrößte Strahlenschaden der Geschichte', *Thüringer Allgemeine*, 29 August 1992, Beilage, 160.4.

[27.] Viertel, *Sankt Urban*, 42.

[28.] Ibid., 33.

[29.] Ibid., 220.

[30] Ibid., 285.

[31.] See Catherine Field, 'Child martyrs of Germany's own Chernobyl', *The Observer*, Sunday 16 May 1993, 20.

13 • Green Thought in Modern Swiss Literature

Jürgen Barkhoff

Empörung durch Landschaften [*From Landscapes to Revolt*] is the programmatic title of a collection of radio essays by the Swiss writer and professor of German literature Adolf Muschg. Inspired by the Romantic landscapes in paintings by Karl Friedrich Schinkel, the central essay which lent its name to the volume as a whole discusses our loss of perception of nature around us in its wholeness and as an entity, and sees this as a consequence of the destructive and alienating effects of our civilization. The essay culminates in an appeal for outrage at this development, an appeal which Muschg says must be 'uncompromising, no longer intimidated by concrete, a protest come hell or high water'.[1] The link Muschg establishes between aesthetic experience and political activity is characteristic of much of the Swiss literature that is the subject of this chapter: literary landscapes, explicitly or implicitly depicting and reacting to today's environmental crisis. They are aimed at creating a sense of loss in view of man's disturbed relationship with nature, reflecting critically on the process leading to this, or at re-establishing a bond between man and his natural environment, in each case with the intention of raising critical consciousness and inspiring practical action.

From the outset Swiss writers have been prominent in the environmental movement. As early as 1977 Otto F. Walter, the most outspoken political writer of the second post-war generation, attacked the 'headlong rush of progress into madness' in a speech at an anti-nuclear power demonstration at Gösgen.[2] In the eighties Franz Hohler, the leading Swiss satirical revue performer, became one of the figureheads of the green movement in Switzerland. And in 1982 E. Y. Meyer published a philosophical essay, the lengthy title of which clearly indicates the concern that inspired it: *Plädoyer – Für die Erhaltung der Vielfalt der Natur beziehungsweise für deren Verteidigung gegen die ihr drohende Vernichtung durch die Einfalt des Menschen* [*Plea for the Retention of Natural Diversity, or for the Defence of the Same against the Threat of Destruction through Human Stupidity*].[3]

These are but a few examples of a commitment to the environment that is very much in line with the self-definition of the role of the writer in present-day Switzerland: as a sensitive critic of society, as an advocate for its neglected or even repressed issues, as a voice disturbing its certainties and undermining its seemingly rock-solid self-confidence. It is not, however, the political stage that was and is the prime arena for such commitment, but rather literature; an idea expressed in Otto F. Walter's plea for 'a literature which is open to the question of what it can still contribute in this latent catastrophe . . . a literature which gets right up close to the catastrophe'.[4] In that sense the looming ecological catastrophes of today are the underlying reason why the environment and man's problematic relationship to nature are featuring prominently in the contemporary literature of German-speaking Switzerland. They feature not only in the so-called 'Öko-Literatur'[5] that identifies directly with the analysis and the political goals of the green movement and describes ecological disasters and man's reaction to them, but also in works that display a more reserved or even sceptical attitude towards the green agenda. In the limited scope of this chapter I intend to discuss some works of contemporary Swiss writers that represent a range of different attitudes towards nature and the ecological question, ranging between scepticism and commitment, anthropocentrism and biocentrism, art and action.

There are several reasons why nature is a theme for Swiss writers in particular. One is certainly the dominant presence of the Alps both as a massive, not to say overwhelming manifestation of nature and as a fragile and endangered ecosystem. Few things have had a greater impact on the way of life, mentality and cultural traditions of the Swiss than their central mountain massif. While in the past the Alps were in many ways a hostile, often overpowering force that, if one was to survive, had to be confronted or won over as an ally, they are today predominantly an idyllic, even awe-inspiring counterpart to a civilization which is perfectly functional, but cold and unexciting.[6] This dependence on the Alps as recreational resort, the relative smallness of the country, its dependence on tourism and its increasingly problematic role (along with Austria) as the main focus of the ever-increasing transit traffic through the Alps: all these factors heighten the awareness of the Swiss that it is precisely their highly industrialized and highly motorized life-style that today poses the main threat to the very heart of their environment.

In this sense Switzerland can well be called the test case which E.Y. Meyer singles it out to be in his *Plädoyer*. It stands paradigmatically for the industrialized nations, in which the growing environmental problems are met by a similarly growing environmental awareness and a sensitivity to the disappearance of nature around us; a trend that is reflected not only in the political and scientific debate, but also in cultural manifestations. And there is an additional reason why Swiss literature in particular should pick up on this topic. Malcolm Pender rightly identifies regionalization and globalization as the two major but conflicting trends in German-speaking Swiss literature since 1945.[7] While the focus on regional belonging that has dominated Swiss writing since the early seventies reflects and counteracts the growing feeling of dislocation and loss of identity, the stress on global interdependence serves not least as a critical objection to the strong and growing isolationist tendencies within Swiss society. The environmental question, at the very heart of which lies the interconnectedness of local causes and global effects, very effectively links the narrow sphere of the *Heimat* to global trends, and thus serves as the strongest reminder that there is no cosy isolation in today's global village.

This link is particularly palpable in *Das sanfte Gesetz* [*The Gentle Law*], the last volume of Silvio Blatter's *Freiamt-Trilogie* [*Freiamt Trilogy*], probably the best and best known example of 'welthaltigen Regionalismus'[8] [regionalism in a world context] of the new, critically conceived type of *Heimatromane* [regional novels]. In it the Chernobyl cloud, the imported nuclear threat, looms over the scenery and action of much of the novel. It highlights one of its central themes, the realization of the growing estrangement between man and nature and the increasingly futile desire to restore a sense of belonging through a rapport with the familiar places and natural scenery of one's *Heimat* [home region].[9]

Such fragile attempts at finding or reasserting one's identity in communication with nature in general, and childhood places in particular, also feature prominently in Otto F. Walter's *Schweizerromane* [Swiss novels].[10] One of them, *Wie wird Beton zu Gras* [*How does Concrete become Grass*] (1979) takes the home-made nuclear threat and the concomitant protest movement as its starting-point. It opens with the Whitsun march of May 1977 in protest at the nuclear power plant in Gösgen, where Walter delivered the speech quoted above, and demonstrates how the intolerant and hostile reaction of the establishment to it has disturbing consequences for the life of the adolescent female protagonist. But although the starting-point and indeed the title of this book seem explicitly 'green', Walter's prime concern, here as always, is not the environment itself. Instead he uses the first ecological movement and its impact on supporters and opponents to demonstrate the alienating effects and repressive authoritarian structures of a capitalist society, obsessed with power and profit.

The possible role of nature in the fight against exploitation and alienation is depicted in more detail in the novel which preceded this one, *Die Verwilderung* [*Decivilization*] of 1977, although nowhere in this book are ecological questions explicitly alluded to. It describes and discusses the attempt of a few young outsiders to establish a small utopian community based on equality, self-determination, and the abolition of all private property. However, this nucleus of an alternative life-style within contemporary Swiss society finally fails because of the distrustful and increasingly hostile reaction of the outside world.

The scenery of this utopian experiment is an old, abandoned clay quarry, the 'Huppergrube'. The symbolic significance of this location works on different levels. Its geographical distance from the town enhances the position of the protagonists as outsiders, their inner distance from society. The fact that it is a ruin of civilization, a natural resource fully exploited and thus abandoned, a scar in an originally idyllic landscape, mirrors the destructive mechanisms of the capitalist society under which nature suffers just as do the protagonists of the novel, and to which they are seeking a more humane alternative. It also aptly reflects that there is no proper place left for a 'pure' utopia, no sphere completely outside existing society, just as there is no unspoiled realm of naturalness, no nature untouched by the forming and deforming influences of civilization. But paradoxically, the fact that the clay quarry has already been fully used as a resource, and is no longer of any interest to its exploiters, means that nature can regain some of its dignity and claim back some of its territory. Walter uses the recurrent image of a female fox, a hunted but proud and independent species, which inhabits the semi-wilderness around the clay-pit. More importantly, in vivid and colourful images Walter shows the protagonists of the novel in a positive interaction with the nature around them. In this interaction they regain strength and reassure themselves that they possess the inner resources needed for their difficult experiment. The book opens with a scene where Rob, the initiator of the commune, dives naked into the flooded clay-pit and resurfaces from its waters, recreating an almost archetypal image of rebirthing through the fluid element. But Walter is anything but naive about this. By beginning his book with a reference to its fictive nature, he highlights the hypothetical, not to say utopian, quality of this scene: 'To create the old clay quarry with these letters on the empty page, and to begin with the 25-year-old . . . '[11] He links such an example of a positive, supportive man-nature relationship to childhood memories of *Heimat*, when another member of the commune recollects a similar youthful dive into the flooded pit as a momentary experience of oneness:

> And suddenly this powerful feeling had been there, a kind of oceanic feeling which had practically inundated him, he had been at one with everything around him, had been a part of it and at the same time he had been happy in quite an extraordinary way.[12]

Such correspondence with nature is also understood as a necessary condition for the inner liberation that is at the heart of the experiment in an alternative life-style. As Leni, Rob's girlfriend, claims, 'At least Rob is already almost free. Rob wants to taste the air and the water.'[13]

But the novel also reflects sceptically whether such a return to nature is possible at all. The solidarity of the group is constantly undermined by conflicts about rivalry and jealousy, mainly in sexual relationships, a sphere of man's inner nature closely linked to the experience of outer nature throughout the book. There are also the various quotations from anthropological treatises that are an integral part of Walter's montage technique. Both elements pose the question of whether the inner nature of the protagonists is already too deformed to allow a breaking-free from the destructive patterns of the possessive and aggressive society in which they grew up. The devastated locality of their experiment, this open wound of outer nature thus also mirrors the wounded inner nature of man and constantly calls into question the fundamental assumption of any utopia, the possibility of a restoration of man's good and sociable nature against the formative forces around them. Though Walter uses his literary landscapes as loci of harmonious correspondence with nature, his narrative technique constantly emphasizes their fictionality and counterbalances them through distancing discursive elements.

A similar strategy to put distance between man and nature is deployed by Max Frisch in his novella *Der Mensch erscheint im Holozän* [*Man Appears in the Holocene*] of 1979. Though Frisch's text, which in an earlier draft was entitled 'Klima' ['Climate'],[14] is more directly motivated by ecological concerns than Walter's, his theme is not the positive side of a closeness to nature without and within, but on the contrary nature's alien and threatening quality. On the surface level the story is a quiet parable about ageing and illness, decay and the human struggle against it. Its only protagonist is the 73-year-old Herr Geiser from Basle. His village in a remote valley of the Ticino, the Italian-speaking part of Switzerland, has been cut off from the outside world by a landslide, caused by a midsummer thunderstorm. For a few days the only road is blocked, food supplies cannot get through and the electricity is interrupted. No ecological catastrophe, but a reminder of the elemental forces of nature, which demonstrates how fragile the ground really is on which the certainties of our civilization rest. Frisch tells his reader in an autopoetological remark how the importance of this today pushes aside his old central theme, the formation of personal identity through our fellow human beings:

> There's nothing left but reading. (Novels are quite unsuitable these days, they are about people in their relationships with themselves and others, about fathers and mothers and daughters ... and about society and so on, as if we were on solid ground, the earth once and for all earth, the sea-level fixed once and for all.)[15]

In the course of the narration this point is picked up again: 'if the ice in the Antarctic melts then New York will be under water, the same goes for Europe, except for the Alps.'[16] But apart from that, there are only indirect references to ecological problems as the reason for such fears that civilization might be wiped out by man-made natural catastrophes. One example is another passage on Herr Geiser's reading preferences: 'Herr Geiser loves non-fiction (Brighter than a Thousand Suns); he has repeatedly read the logbook of Robert Scott, who froze to death at the South Pole.'[17] *Brighter than a Thousand Suns (Heller als Tausend Sonnen)* is the title of a book in which the futurologist Robert Jungk in the early sixties praised man's ability to create a new paradise by using and successfully handling nuclear energy, the most powerful natural force. While Frisch was working on *Der Mensch erscheint im Holozän*, another book by Robert Jungk created a great sensation. *Der Atom Staat [The Nuclear State]* of 1977 was a radical criticism of uncontrollable nuclear technology and a total revision of Jungk's previous optimism. But Frisch consciously abstains from directly alluding to this contemporary work. His reference to the individual fate of the failed polar explorer Scott is much more effective in pointing to the historical dimension of his scepticism as to whether the hostile aspects of nature can be overcome. In *Der Mensch erscheint im Holozän* Frisch is neither concerned with day-to-day environmental politics nor with a general criticism of our techno-logical age as put forward in his *Homo faber [Homo faber]*. Instead he uses an exemplary individual case to demonstrate the ephemeral status of the human species in the context of the overwhelming history of nature. For that purpose he presents Herr Geiser in a situation where he is stripped of the amenities and securities of civi-lization and is thus more than usually exposed to nature. As for Walter, nature here means not only and not primarily outer nature. The extraordinary circumstances in which the protagonist finds himself confront him with his own nature, his ailing physical and deteriorating intellectual capacities.

Both the landslide (the threat from outside) and the fragility of his body (the threat from inside) mobilize his resistance. First his intel-lectual reserves come into play: he starts to cut out relevant quotations and passages from his small library, an encyclopaedia and the Bible, and sticks them to the walls of his living room. With the help of this collage he tries to objectify his situation and to free himself from nature's suffocating embrace by means of historic memory and intellectual overview. Frisch here adds an essayistic and philosophical dimension. He expresses it in the text's aesthetic structure by representing these excerpts with a grey background and in their original typeface. Thus Herr Geiser's individual fight against decay and oblivion represents the attempt of the species to activate cultural memory against its own transience. When he collates geological data about the origin of the Alps, quotes from the biblical creation story and places both beside fragments of the cultural history of the Ticino that illustrate the impact of the mountains on the way of life of its inhabitants, he aims to rise above natural history by defining man's place within it. A quotation from the entry 'Mensch, homo, anthropos' in his encyclopaedia stresses this ability to set oneself off against the world of objects as an intellectually independent subject: 'This ability to distance oneself from the world is the prerequisite for seizing power over it, and thus also for man's privileged position.'[18] But within the narration the reassurance that is achieved with the help of history, science and philosophy remains highly ambivalent and finally futile, only stressing that nature is in principle alien and inaccessible. A gust of wind finally destroys Geiser's order of knowledge on the wall, and the narrator comments: 'All the notes, whether on the wall or on the carpet, can go. What does holocene mean? Nature needs no names. Herr Geiser knows that. The rocks don't need his memory.'[19]

In addition to the written word as medium of reflection and memory, Frisch follows a Faustian tradition in establishing action as the second device to escape the otherness of nature. One morning Herr Geiser sets off to walk over a mountain pass into the neighbouring valley to catch the bus to the next town, i.e. to civilization. It is a long and arduous expedition into hostile and untamed nature, made additionally difficult by age, fog and rain. When after eight hours and a lot of wandering about in unfamiliar territory he hears the bells of the church beside the bus-stop in the not-too-far distance, he knows that he has done it – and returns. Within the story this is a turning-point of the kind that characterizes the classical novella: an 'unerhörte Begebenheit' [unheard-of incident] which changes the life of the protagonist and reverses the perspective of the text. Herr Geiser's attempt to break free did not fail; in realizing that there is no escaping from the dependence on nature outside and within he instead decided not to go through with it. Paradoxically this – albeit resigned – approval of his own naturalness is an act of self-determination, an act of free will in a Kantian sense. Consequently, the next day Herr Geiser suffers a small stroke that speeds up and seals his regression into the natural state. Within the context of the narration this is motivated by the physical strains of the previous day; within the logic of Frisch's philosophical argument it is an inevitable consequence, foreboding the fate of humankind.

Although environmental concerns were a major motivation for this text, it transcends topical green issues and integrates them into a wider perspective. Characteristically, and in contrast to the following texts, there is no attempt to put blame on our civilization for nature's hostile reaction. Hence a work of literature that has predominantly been read as a piece about ageing[20] can indeed be seen as an example of the increasingly detached but nevertheless all but dispassionate wisdom of old age.

Like Frisch's novella, Walther Kauer's short novel *Spätholz* [*Late Wood*] (1976) is set in a remote valley in the Ticino and has as its sole protagonist an ageing man who is increasingly alien in his once familiar surroundings. It too has a landslide as its central event. Although in *Spätholz* the landslide does not initiate the action, but is its culmination, the natural catastrophe here too serves as a symbol of the slipping away of once rock-solid certainties. But despite these similarities, the aesthetic approach of the two texts to their topic is distinctly different. While Frisch's narrative technique constantly establishes distance between the reader and its theme, and thus forces us to take on an ambivalent and critical attitude, Kauer creates little or no distance from his protagonist but draws the reader into his story and its value judgements. We witness the last twenty-four hours in the life of the old mountain farmer Rocco. He is staying awake through the night, determined to prevent the felling of an old walnut tree in his grounds, which in a magic ritual was planted with his placenta underneath on the day of his birth. It is to be felled because it impedes the lake-view of his new neighbour, a retired German industrialist. Somewhat black and white as this confrontation between old and new may be, it adequately reflects the dramatic social changes the Ticino underwent in this century. Once the poorest region of Switzerland, its outstanding natural beauty destined it to become the preferred holiday resort for the rich and the famous, a development that threatens the identity and the ecological balance of the region. In this context Rocco stands for a dying way of life, based on hard work, modesty, religion, strong traditions and – most importantly for our theme – a closeness to nature. Much of the book consists of Rocco's memories of this traditional way of life, recollections of the gradual changes during his lifetime and his resistance to them. The latter was in large part a fight against ecological blindness, first in an attempt to prevent a huge reservoir from being built, and after that in warnings that the mountains above the village would not withstand the pressure of its volume of water. So, as in all the books of the late seventies mentioned here, the ecological topic is part of a wider concern; in this case the loss of understanding for nature is linked to a general loss of tradition. But Kauer's message is much more directly 'green' than was Frisch's or Walter's. The younger generation now in charge that wants to get rid of Rocco's walnut tree for the sake of tourism and foreign money is the same one that neglects the *Bannwald* [protective forest] above the village and replaces it with inefficient concrete protection walls. The day the walnut tree is felled, the village (and Rocco) are buried under a massive landslide. In his last hours Rocco, instead of keeping to his original plan to prevent the felling of his tree, reads the danger signs of nature and tries to warn his fellow villagers. But they will not listen: 'It had nothing to do with fate if people did their utmost to destroy the Terzone ... If they forgot the mountain. Or worse still: if they ignored its unmistakable warning signs.'[21] The catastrophic landslide is understood as nature's revenge on man, for his disrespect and arrogance, for his loss of perceptiveness towards nature. Rocco is portrayed as a representative of a bygone age, long ago initiated into communication with nature by the old folk-healer and heathen magician of the village. His correspondence with nature is also symbolized by the way his fate and that of the village as a whole are linked to that of the walnut tree.

For Kauer, Rocco's attentive and caring alliance with nature is irretrievably lost; his book is basically a statement of resignation and nostalgia. The same notion of an animate nature, communicating with men, is also at the centre of *Der neue Berg* [*The New Mountain*] (1989), the first and so far only novel of the well-known Swiss satirist Franz Hohler.[22] But for Hohler such an empathetic relationship with nature takes on a much more active role in the fight against apathy and blindness. In that sense *Der neue Berg* is the most explicitly 'green' of the texts discussed here. The novel culminates in a virtually impossible event, the eruption of a volcano near Zurich. The book relates the events preceding this outbreak on Whit Sunday 1988: the first, hardly noticeable cracks on the surface of a Celtic tumulus, the later scene of the eruption, a few slight but foreboding pre-shocks and the growing but inadequate public reaction to it. On its final page the volcano is referred to as follows:

> this main fire could not be extinguished, since it came out of the depths of a planet at the core of which an eternal fire raged ... now some force had brought this fire to the surface, here, in our midst, here, where this planet is violated daily.[23]

What has implicitly governed the story throughout becomes explicit at the end: its author wants us to view this natural catastrophe as a revolt of the wounded and violated earth, which finally responds with its own elemental destructive force to the destructive tendencies of civilization. The long account of the preceding events serves to demonstrate that this is nature's reaction to what man does to her: it is an act of revenge.

Before the Enlightenment, volcanic eruptions of course were always interpreted as a destructive fire from above, sent by God as warning or as punishment. But this fire comes from below; it is an attack from the depths of our planet. In stressing this perspective Hohler uses another pre-modern idea, taken from myth and hermetism: that of the earth as Gaia, as a living being. The anachronism of this animistic concept, rejected by philosophy and science, is expressed in the surrealistic quality of its central event.[24]

As the novel approaches its climax it depicts a vivid and complex kaleidoscope of Switzerland during the eighties. Narrated in realistic detail and with a lively sense of humour, this has several functions. Without ever establishing a direct causal link between specific ecological wrongdoing and the eruption, it portrays in passing our ecological sins large and small, as well as the feeble attempts to counteract them: the devastation of the landscape through roads for ever more cars set beside the pleasures of cycling, the large-scale wastage of water and the use of environmentally friendly washing-powders, the eradication of the tropical rain forest for new pastures to satisfy European beef demand and vegetarian cuisine, the hole in the ozone layer and local protest groups.

Another purpose is to demonstrate the total inadequacy of our normal ways of dealing with mostly unspecific but constantly looming ecological dangers.[25] This is mainly exemplified by the contrastive reaction of two antagonistic groups: the town officials around the mayor Niederer, and the worried citizens around Roland Steinmann and the alternative grouping in the town parliament, the 'Fresh Wind'. The main conflict between these two sides is exactly the one governing current political debates about ecological risks. While the worried citizens accuse the officials of playing down real dangers, the latter claim that the protesters are causing unnecessary panic and hysteria because of the statistical improbability of the event announcing itself.

An example of the blindness of the latter is the first episode involving Niederer. He is introduced as a patient suffering from an inexplicable allergic skin irritation. He treats it as a defect of his body machine that has little to do with him, and receives sixteen penicillin injections, which are, needless to say, ineffective. It is easy to see that here, on a small scale and within the body of an individual, the same is happening as the novel later enacts on a large scale. Nature takes its small, private revenge with Niederer, so to speak. As a typical effect of environmental pollution his illness is the microcosmic equivalent of the global illness of the macrocosmos, earth. So, early on in the book Hohler introduces the mutual dependence between the body of the earth and man's physical nature; it is a concept equally central to today's ecological debates and to the archaic idea of the planet as a living being.

In contrast to Niederer, the worried citizens show an intuitive grasp of such correspondences; they are able to receive and interpret the subtle warning signals that the earth is sending. For Hohler the first admonishers are the children, as they live in an animistic world anyway. An example is the nine-year-old Fabian who, since the first pre-shocks, keeps having nightmares that his hands are growing enormous and will explode. In his first play session with a child psychologist he enacts a volcanic eruption, and in the second a large blaze. When asked the reason for the latter, he answers cryptically: 'the earth opened up.'[26] 'Deeprooted fears', the therapist jots down after the session, not realizing that this is not merely the diagnosis for the boy's problems, but for those of the earth itself. But there is yet another level of diagnostic value in this scene: one anthropological distinction between man and animal is the anatomical build of our hands. Our thumb enables us to use tools, to be a craftsman, a technician, a *homo faber*. The uncontrollable explosion of our technical capabilities lies at the root of the present ecological crisis.

In various scenes the novel tackles the question – one which is central to its ecological message – of whether the dreams, fantasies and images through which a subconscious awareness of the coming danger expresses itself, are anything more than subjective projections. One such is Roland Steinmann's last visit to the Celtic tumulus, when hissing sulphuric fumes are already emanating from its gaping cracks: 'Was he mistaken or did he hear the earth breathing? Or was it his own breathing? After he got up he had to rub his eyes again. Was he mistaken or was the fog burning him?'[27] He certainly does not fool himself in this last respect. That means not only that he is right in the first respect too, but also that the correct understanding of the state the earth is in depends on this animistic mode of perception.

Within the story such intuition motivates action, and enables Roland Steinmann and his friends to persuade many people to flee, rescuing them from a disaster that kills over 20,000 people. On the evening preceding the catastrophe Roland Steinmann argues with the head of the town's emergency services that 'the danger must be gauged in terms of our own fear. We must trust our own feelings'. To this the latter replies that this is 'utter nonsense ... Psychobabble and lack of leadership, that's all they are, our feelings.'[28] The insistence on the relevance of subjective feelings and the notion of being directly affected of course play a major role in the green movement's understanding of itself. And so does – at least for the sizeable part of it influenced by New Age philosophy – the concept of the earth as a living being, communicating with and reacting to its inhabitants. One only needs to mention the Gaia hypothesis of the British scientist James Lovelock. It is certainly no coincidence that Hohler repeatedly links the forebodings of the violent reaction of the earth to the Celts, a people that at present enjoys a high, even if for lack of sound evidence somewhat vague, esteem in esoteric circles for their alleged closeness and kindness to nature. But that does not mean that Hohler should be seen as an apostle of the esoteric return of the Celts or as a biocentrist. Such an identification is counteracted by the satirical and surrealistic qualities of his book. But Hohler certainly has sympathy for the notion that we should not treat the earth as a dead object that has nothing to do with us, and his fictional images of a living earth and his anthropomorphic metaphors[29] are strong reminders that, as Joachim Ritter, Hans-Robert Jauss and more recently Gernot and

Hartmut Böhme have pointed out, the aesthetic sphere today is the last realm to express an experience of nature that goes beyond a detached scientific or instrumental technical attitude.[30] In demonstrating how such aesthetic perception can translate into political action, Hohler's novel is a fine and rare example of how, without being overly didactic, eco-literature can depict and promote the *Empörung durch Landschaften* Adolf Muschg is calling for.

But such an optimistic position does not of course remain uncontradicted in contemporary Swiss writing. A highly original and skilful example of this counter-tendency is Hermann Burger's *Die Wasserfallfinsternis von Badgastein* [*The Eclipse of the Badgastein Waterfall*], a short novella that won the prestigious Ingeborg Bachmann prize in Klagenfurt in 1985. What makes it an especially effective repudiation of any claim that art can lead a way into nature, is the fact that it comes in the guise of a piece of *Öko-Literatur*. There are many thematic parallels to Kauer and Hohler. Again a natural catastrophe, including an earthquake (and the massive emission of highly concentrated natural radioactivity), is to be understood as a 'protest against the insane exploitation' of nature, against 'humankind's plundering of its resources'.[31] This interpretation of the story's main event, the drying-up of the central waterfall of the famous Austrian thermal spa Badgastein, is communicated to the first-person narrator by nature itself, which seems to be another parallel. Carlo Schusterfleck, the night porter of a hotel in Badgastein, seems in two respects predestined for such a privileged rapport with nature. In the Romantic tradition he is seeking physical health and artistic inspiration alike. He is not only suffering severely from *morbus bechterev*, for which the treatments in this spa are specifically beneficial; he is also a Schubert enthusiast who hopes to find in the sounds of the cascades of the waterfall the central motif of Schubert's missing Gastein symphony.

> . . . I was always obsessed with the idea that if only it were possible to nestle the twisted vault of ribs which was my Bechterev hump into this echo of nature, as though into her primal sound, then the advancing sclerosis might be stopped . . . the theme of the missing symphony would then resound.[32]

And indeed, both predispositions allow him to understand the language of nature. Wandering around at night in search of healing and inspiration, he is the first and only one to discover that the waterfall has dried up. Paradoxically it is the now silent waterfall that speaks to him and reveals in his testament that this is a deliberate 'suicide as a *coup d'état* of nature'.[33] It also reveals to Schusterfleck the central motif of Schubert's Gastein symphony and, furthermore, discloses that it was not inspired by the cascades of the waterfall, but by the rhythm of a Grillparzer poem inscribed on the wall of one of the hot springs. That is nature's message to man: that art is inspired by art, and not by nature.[34] So the Romantic philosophy of a *Natursprache* [language of nature] communicable to man which Hohler cites affirmatively is used by Burger ironically to reject its very claims.[35] Additionally, this suicide of a waterfall as an act of ecological protest must, of course, be read as a variation of Burger's dominant theme of self-extinction, this time projecting his personal death wish on to nature as a whole.[36] One paragraph in Burger's *Tractatus logico-suicidalis*, a justification of the suicide he committed shortly after its publication, suggests the same radical projection of subjectivity: 'Given the nuclear and ecological threat to the world – the looming omnicide – suicide as a solution is an artistic-revolutionary act: the person who commits suicide anticipates – *pars pro toto* . . . – that which in all likelihood will happen again, sooner or later.'[37] Finally, Burger's text is written in the highly artificial language so characteristic of him, constantly transgressing the familiarities of a more 'natural' conventional language. And his text is structured in five movements like a symphony, so itself not imitating nature but art.

Burger is not the only one who reintroduces the Romantic concept of *Naturmusik* [music of nature] to show its inappropriateness. In Gertrud Leutenegger's novel *Kontinent* (1985) the first-person narrator is commissioned by an aluminium factory to record the sounds of nature of the very Valaisian valley which it has devastated ecologically. They are to be integrated into a piece of *Naturmusik* which 'had penetrated through all of nature and its raped elements so unconditionally that it could become one with the roaring on the inside'.[38] But the novel exposes this notion of music as a harmonizing force mediating between man and nature as sheer ideology. The composition including the sounds of nature is for a record to celebrate the jubilee of the factory. The closeness of art and nature here just means that both can be exploited with equal cynicism, and by the same people. But Leutenegger's haunting descriptions of environmental devastation as well as, on a formal level, the surrealist and Kafkaesque elements of the novel also point to the remaining functions of art: at least to disclose and undermine such ideological instrumentalization and to create melancholy and anger.[39]

It is perhaps not surprising that E. Y. Meyers's attempt to portray a similar kind of *Naturmusik* in a truly reconciliatory role did not materialize. For years the author worked on a novel called *Das Naturtheater* [*The Nature Theatre*]. It centred on a day-long musical performance in a huge nature arena near Trieste, in which the rhythms and sounds of nature, recorded and transformed by computers, were meant to blend harmoniously with the whole musical tradition of the west in a similar way. The use of the most advanced technological devices to foster such an ambitious *Naturversöhnungsprojekt*[40] [project of reconciliation with nature] is perhaps as indicative of its aporia as is the fact that the novel was never finished.

Art is most certainly overburdened with the task of reconciling man and nature in the present ecological crisis. Even to suggest that they should is a gross misrepresentation of their role in society. Summarizing, one might instead point to two major functions of the aesthetic experience created by and conveyed through literature. If one looks at the texts by Frisch, Burger and Leutenegger, highly artistic narrative devices and a mannered literary language highlight the alienation of man from nature. But Walter, Kauer and Hohler suggest just as Muschg did that we need and can use the aesthetic experience to empathize with nature. In that perspective all the landslides and earthquakes in contemporary Swiss writing serve as warning signals that indicate to what degree such an understanding of nature has already been lost and what the consequences will be if we fail to recultivate it.

Notes

[1.] Adolf Muschg, *Empörung durch Landschaften. Vernünftige Drohreden* (Frankfurt am Main, 1988), 15–18, here 18. Translations of this text and others throughout by Nicola Creighton.

[2.] Otto F. Walter, 'Das Veto des Volkes. Erste Gösgener Rede', in his *Gegenwort. Aufsätze, Reden, Begegnungen* (Zurich, 1988), 60.

3. E. Y. Meyer, *Plädoyer – Für die Erhaltung der Vielfalt der Natur beziehungsweise für deren Verteidigung gegen die ihr drohende Vernichtung durch die Einfalt des Menschen* (Frankfurt am Main, 1982).

4. Otto F. Walter, 'Literatur in der Zeit des Blindflugs' (1978), in Walter, *Gegenwort*, 65.

5. For an overview of *Öko-Literatur* see Reinhilde Wiegmann, 'Graue Gegenwart – Aufgehellte Zukunft. Literarische Texte im Umfeld der Grünen', in Jost Hermand and Hubert Müller (eds.), *Öko-Kunst. Zur Ästhetik der Grünen* (Hamburg, 1989), 125–48. Swiss authors (but not those covered in this chapter) are discussed in Manfred Gsteiger, 'Zeitgenössische Schriftsteller im Kampf für die Umwelt', in Manfred Schmeling (ed.), *Funktion und Funktionswandel der Literatur im Geistes- und Gesellschaftsleben. Akten des Internationalen Symposiums Saarbrücken 1987* (Berne and Frankfurt am Main, 1989), 101–12. See also ch.5, 'Thema Natur' in Yvonne Denise-Köchli, *Themen in der neueren schweizerischen Literatur* (Berne and Frankfurt am Main, 1983), 156–71.

6. For the cultural-historical background of this development see Clemens Alexander Wimmer, 'Die Alpen. Vom Garten Europas zum Stadion Europas', in Jost Hermand (ed.), *Mit den Bäumen sterben die Menschen. Zur Kulturgeschichte der Ökologie* (Cologne, Weimar and Vienna, 1993), 81–117.

7. Malcolm Pender, 'Trends in writing in German-speaking Switzerland since 1945', in Michael Butler and Malcolm Pender (eds.), *Rejection and Emancipation: Writing in German-Speaking Switzerland 1945–1991* (New York, Oxford, 1991), 24–38, here 24.

8. Martin Zingg, 'Besuch in der Schweiz', in Klaus Briegleb and Sigrid Weigel (eds.), *Hansers Sozialgeschichte der Literatur, 12. Gegenwartsliteratur seit 1968* (Munich, 1992), 643–66, here 645.

9. Silvio Blatter's novels *Zunehmendes Heimweh* (1978), *Kein schöner Land* (1983) and *Das sanfte Gesetz* (1988) would deserve a more thorough discussion in the context of such an overview. But as they have just recently been treated with regard to their 'green' implications, I content myself with a reference to J. H. Reid, 'Silvio Blatters Romantrilogie *Tage im Freiamt*: Der Öko-Roman zwischen Heinrich Böll und Adalbert Stifter', in Axel Goodbody (ed.), *Literatur und Ökologie. Der literarische Beitrag zur ökologischen Debatte in den siebziger und achtziger Jahren* (Amsterdam, 1997) (forthcoming).

10. For a useful English introduction to Walter's *œuvre* see Ian Hilton, 'Otto F. Walter: literature and the strategies of revolt', in Butler and Pender (eds.), *Rejection and Emancipation*, 59–73.

11. Otto F. Walter, *Die Verwilderung* (Reinbek bei Hamburg, 1981), 7.

12. Ibid., 267f.

13. Ibid., 38.

14. See Walter Schmitz, *Max Frisch: Das Spätwerk (1962–1982)* (Tübingen, 1985), 140, 143.

15. Max Frisch, 'Der Mensch erscheint im Holozän', in his *Gesammelte Werke in zeitlicher Folge. Jubiläumsausgabe in sieben Bänden*, 7 *(1976–1985)*, ed. Hans Mayer in collaboration with Walter Schmitz (Frankfurt am Main, 1986), 205–99, here 211f. It is worth noting that the personal interrelationship between fathers, mothers and daughters is exactly that of Frisch's novel *Homo faber*, his most outspoken criticism of the scientific and technological age.

16. Ibid., 271.

17. Ibid., 212.

18. Ibid., 271.

19. Ibid., 296.

20. An exception is the convincing 'green' reading by Gerhard F. Probst, 'The old man and the rain: man in the holocene', in G. F. P. and Jay F. Bodine (eds.), *Perspectives on Max Frisch* (Kentucky, 1982), 166–73. See also Wulf Koepke, *Understanding Max Frisch* (Columbia, 1991), 108–13.

21. Walther Kauer, *Spätholz* (Zurich and Cologne, 1976), 226.

22. For some general information about Hohler and his various roles as satirist, revue performer, author of children's books, screenplays and short stories see Michael Bauer and Klaus Siblewski (eds.), *Franz Hohler. Texte, Daten, Bilder* (Hamburg and Zurich, 1993).

23. Franz Hohler, *Der neue Berg* (Munich, 1993), 433.

24. Hohler had previously expressed the same idea of nature taking revenge and regaining lost territory in his surrealistic short story *Die Rückeroberung* (Hamburg, 1982), in which wild beasts and thick vegetation invade and take over the city of Zurich. For a further example of Hohler's animistic perception of nature see his account of a mountain tour in *Museumsbesuche. Schweizer Schriftsteller schreiben zu Bildern der Stiftung Oskar Reinhart* (Frankfurt am Main and Leipzig, 1993), 15–19.

25. In that sense the volcanic eruption can adequately be read as a metaphor for the greatest imaginable man-made environmental disaster: the meltdown of a nuclear power plant.

26. Hohler, *Der neue Berg*, 291.

27. Ibid., 378.

28. Ibid., 417.

29. See ibid., 42, 57, 61f., 79f., 86f., 115, 169, 202, 212, 358f., 376, 410.

30. See Joachim Ritter, 'Landschaft. Zur Funktion des Ästhetischen in der modernen Welt', in his *Subjektivität* (Frankfurt am Main, 1974), 141–63; Hans Robert Jauss, 'Aisthesis und Naturerfahrung', in Jörg Zimmermann (ed.), *Das Naturbild des Menschen* (Munich, 1982), 155–82; Gernot Böhme, *Für eine ökologische Naturästhetik* (Frankfurt am Main, 1989); Hartmut Böhme, 'Aussichten einer ästhetischen Theorie der Natur', in Jörg

Huber (ed.), *Wahrnehmung von Gegenwart. Interventionen* (Basle and Frankfurt am Main, 1992), 31–53.

[31.] Hermann Burger, 'Die Wasserfallfinsternis von Badgastein. Ein Hydrotestament in fünf Sätzen', in his *Blankenburg. Erzählungen* (Frankfurt am Main, 1986), 161–79, here 174.

[32.] Ibid., 169.

[33.] Ibid., 173.

[34.] For a more detailed discussion of this see the convincing analysis of Monika Großpietsch, *Zwischen Arena und Totenacker. Kunst und Selbstauflösung in Leben und Werk Hermann Burgers* (Würzburg, 1994), 190–4.

[35.] For modern reverberations of *Natursprache* see Axel Goodbody, *Natursprache. Ein Konzept der Romantik in der modernen Naturlyrik* (Kiel, 1984).

[36.] See, for this notion of self-extinction, John J. White, 'Hermann Burger: "Die allmähliche Verfertigung des Todes beim Schreiben"' in Butler and Pender (eds.), *Rejection and Emancipation*, 184–201. Another example of the projection of subjectivity on to nature is Burger's short story 'Zentgraf im Gebirg oder das Erdbeben zu Soglio', from his collection *Diabelli. Erzählungen* (1979). In it the narrator attributes an earthquake in southern Germany to the death of the eccentric Zentgraf in the Swiss Alpine village of Soglio, where he quite unscientifically locates the epicentre of the earthquake.

[37.] Hermann Burger, *Tractatus logico-suicidalis* (Frankfurt am Main, 1988), no.243, 60.

[38.] Gertrud Leutenegger, *Kontinent* (Frankfurt am Main, 1985), 38.

[39.] For a more detailed discussion of the ecological implications of Leutenegger's work see Elizabeth Boa, 'Gertrud Leutenegger', in Butler and Pender (eds.), *Rejection and Emancipation*, 202–21, esp. 211–14. See also Ann Marie Rasmussen, 'Women and literature in German-speaking Switzerland: tendencies in the 1980s', in Mona Knapp and Gerd Labroisse (eds.), *Frauen-Fragen in der deutschsprachigen Literatur seit 1945* (Amsterdam, 1989), 159–82, esp. 180f.

[40.] That is the telling title of a short excerpt from the novel, published in *Schweizer Monatshefte,* 67 (1987), 831–4.

Part IV
Art and Popular Culture

14 • Joseph Beuys's Eco-aesthetics

Frank Finlay

Introduction

Of the many writers, artists, and other cultural producers who lent their fame, oratory and ideas to the modern ecological movement in Germany, Joseph Beuys is undoubtedly one of the most influential. Beuys, who is considered by many critics to be the pre-eminent artist of the post-war era in Europe, was involved actively in developments that culminated at the end of the 1970s in the emergence of 'alternative' or 'green' politics.

In this chapter, I shall first outline briefly the practical contribution made by Beuys to the green movement before concentrating on a discussion of the ideas central to his anthropological and aesthetic credo. In so doing, I shall refer to salient aspects of his biography as well as to some of his artistic works, and compare his views with those of another prominent supporter of the ecological movement, his friend and confidant, the Nobel laureate Heinrich Böll. In a concluding section, I shall offer a historical perspective on Beuys's aesthetics by identifying some of the intellectual traditions on which he draws.

Demanding an alternative

Joseph Beuys was one of the co-founders of the German Green Party, and in many ways the party's emergence can be regarded as the apogee of his own direct political agitation, which had its origins in the Studentenbewegung [Student Protest Movement] of the 1960s. As a professor at the Staatliche Kunstakademie, the State Academy of Art in Düsseldorf, he was the catalyst for a number of important developments, as the following examples illustrate. Within weeks of the shooting of the student, Benno Ohnesorg, during demonstrations in June 1967 against the visit to Berlin of the Shah of Iran – an event generally regarded as having radicalized the Student Protest Movement – Beuys founded the Deutsche Studentenpartei [German Student Party], which shared many of the concerns of the later ecological and peace groups, not least on the issue of multilateral disarmament.

Beuys soon sought to extend his influence beyond the confines of student politics by setting up a number of often short-lived groups, such as the Organisation für direkte Demokratie [Organization for Direct Democracy], which was based on the conviction that a broader spectrum of people had to be politically active at grass-roots level in order to achieve the radical change in society that he so urgently demanded. In particular, Beuys believed that the established parliamentary parties were no longer representative of the vast majority of the German people, and at one stage he even called for a boycott of the federal elections as a means of breaking what he referred to as the 'dictatorship' of the parties.

In many ways, Beuys anticipates some of the early slogans of Green Party activists like Petra Kelly with her call for the Greens to be an 'anti-party party'.[1] His early championing of 'direct democracy' accompanied by demands for plebiscites to decide important issues of the day was, therefore, undoubtedly a major contribution to the green debate. Similarly, the threat to the environment was another pervading theme of the discussions, seminars and teach-ins that were a feature of these groups' activities. One of Beuys's earliest environmental 'happenings', for example, took place in 1971 in Düsseldorf, under the aegis of his Organization for Direct Democracy, and had as its title an exhortation to overcome both the dictatorship of the political parties and save the forest ('Überwindet die Parteiendiktatur. Rettet den Wald!'). Other political organizations were to follow, such as his Aktionsgemeinschaft unabhängiger Deutscher [Action Community of Independent Germans], for whom Beuys ran as a candidate in the elections of 1976, polling some 600 votes. Almost all of his activities at this time were deployed with the innate skill of a supreme self-publicist in order to propagate his political ideas.

When, in March 1979, the Greens were founded as a 'sonstige politische Vereinigung' [other political association], under the rules for participation in the elections for the European Parliament, it was inevitable that Beuys should run – albeit unsuccessfully – as a candidate. It was in these early days of the party that Beuys enjoyed his greatest influence on its policy-making. This emerges with abundant clarity from his seminal text written for the *documenta 7* entitled 'Aufruf zur Alternative' ['Demanding an Alternative'], first published in the *Frankfurter Rundschau* on 23 December 1978. Beuys's 'manifesto of comprehensive cultural criticism'[2] not only summarizes his major concerns but also articulates many of the positions which were to help the Greens delineate the contours of their socio-ecological programme.[3] Specific reference is made to the threat of atomic destruction, and the malfunctioning of both western private capitalism and eastern state-monopoly capitalism, which negate human needs and result in the alienation of the individual from the Self and the natural world. Beuys identifies this alienation as the major factor behind the crisis in the ecosystem, and he calls for a new politics committed to the restoration of freedom, creativity, solidarity, and mutual affinity.[4]

Many of Beuys's public statements can be regarded as an attempt to formulate a radical alternative to the desolation of life in the second half of the twentieth century, and they cover territory familiar in the main thrust of modern green thought. His most original contribution, however, was to allot to art a political function as a means towards a better and more optimistic society; one in which the individual can fulfil his or her human potential, and recover a state of harmony with the social and natural world. It is to the main tenets of what can be termed his 'eco-aesthetics' that I should now like to turn.

'Social sculpture'

Joseph Beuys's advocacy of art possesses a certain logic when viewed in the context of his own biography. Beuys did not become a professional artist until his early forties, and then only after he had suffered a number of chronic depressions and much physical and emotional anguish. The event which had arguably the most profound effect on Beuys occurred during his wartime service (for which he volunteered enthusiastically), and which saw him become a combat pilot in the Luftwaffe. In 1943 he was shot down over the Crimea, and his life was saved by Tartar tribesmen, who covered his almost fatally wounded body in fat and felt, to keep him warm.

When Beuys became an artist and attempted to symbolize in his work the healing and comforting potential of art, he drew time and again on these and other emblems of protection and nourishment. Although, for his devotees, Beuys's wartime suffering has become 'part of the hagiography of modern art',[5] there is little reason to doubt his own avowal that these experiences unleashed a period of rumination and reorientation. The most important decision he made was to reject science in favour of art. From his childhood, Beuys had regarded science as his true vocation, but he now came to realize that the positivist materialist methodology of science was: 'a restriction on my specific ability and I decided to try it another way'.[6] Following a long period of recuperation after the war, Beuys enrolled at the Staatliche Kunstakademie, Düsseldorf, in 1947.

In the next five years Beuys received a rigorous and classical training as a painter and sculptor under Josef Enseling and Ewald Mataré. He became Mataré's master-pupil between 1952 and 1954, and contributed a great deal to the many commissions which the latter received in the phase of post-war reconstruction, including a bronze door for Cologne cathedral. By the mid-1950s it appeared that Beuys was on the threshold of a very promising career. It was during this time, however, that he realized to his great disappointment and frustration that rigid ideas of specialization were not unique to science; concepts of art could be equally restrictive. In particular, his own experience in the Academy suggested to him that traditional art had been reduced to high-achievement artistry on the part of a number of individuals of recognized genius; a view which ignored the inherent creative potential of every individual. Moreover, art works had become reified by the 'culture industry', and their emancipatory and utopian potential had been neutralized in a world governed by the exclusive dictates of technological rationality.[7] This realization plunged him, in the mid-1950s, into a further, and quite literally life-threatening, crisis. However, with the material help and emotional support of his lifelong patrons, the van der Grinten brothers, he emerged at the end of the decade reinvigorated and convinced that nothing short of a conceptual revolution was required for art to regain its substantive anthropological and social function.

It was from this time onwards that Beuys began to focus on the centrality of art to both experience and regenerative change. The essay 'I am Searching for Field Character' is an important document in this context, not least because it offers a concise summary of his main theoretical concerns. Beuys speaks of the need for a radical widening of the definition of art so that it may reassert itself as an agent of change: 'Only art is capable of dismantling the repressive effects of a senile social system that continues to totter along the deathline: to dismantle in order to build A SOCIAL ORGANISM AS A WORK OF ART.' This is what Beuys refers to as 'social sculpture', which, in his view, will only materialize 'when every human becomes a creator, a sculptor or architect of the social organism'. Beuys is by no means advocating that each of us should abandon our current activities and take up painting; rather he conceives of all social action as a living sculpture; it is an anthropological understanding of sculpture as being related to the social body, and to the lives and abilities of all social subjects:

> Every human being is an artist who – from his state of freedom – the position of freedom that he experiences at first hand – learns to determine the other positions in the TOTAL ART WORK OF THE FUTURE SOCIAL ORDER. Self-determination and participation in the cultural sphere (freedom); in the structuring of laws (democracy) and in the sphere of economics (socialism). Self-administration and decentralization occurs (threefold structure).[8]

Thus, power for change lies in the hands of the individual, and in the transforming force of creativity and imagination, which can be expressed in any medium or subject matter. Beuys's answer to the straitjacket of traditional concepts of art, therefore, was to extend the notion of 'artist' to embrace every individual irrespective of their profession, so that 'art' can refer to all kinds of activity within the overall 'sculpture' of social and political life. Similarly, he hoped to liberate art from its essentially decorative function in the museum and to 'expand' it into life.

All of Beuys's own activities as an artist sought to recreate and reactivate human creativity as a cognitive force, and many of his works deal to a greater or lesser extent with the impact of the ecological crisis. One of his most famous and impressive is *Das Rudel* [*The Pack*] of 1969. This installation consists of twenty sledges pouring out from the rear of a Volkswagen minibus, each carrying its own survival kit of quintessential Beuysian materials: felt (for protection), a ration of fat (for food), and an electric torch (for orientation). The implication, according to Beuys, is that the modern technological world is incapable of dealing with the state of crisis which it has produced. As a result, a more primitive and direct means must be found to ensure our survival.[9]

In his paintings, sculptures and drawings, he attempted frequently to embody a world unsullied by some of the deformations of civilization. Many of his sketches, for example, depict animals and insects, typically hares, stags and bees, as unalienated creatures at one with creation. Beuys's concern with spiritual healing accounts not only for his preoccupation with depictions of Christ in much of his early work, but also explains his interest in shamanism,[10] which he regarded as an attempt to restore contact between the material and spiritual world. There is a clear shamanistic tendency in his actions of the 1960s, like the famous *How to Explain Pictures to a Dead Hare*, and in his imbuing of conventionally repellent materials with symbolic meaning.[11]

Without doubt Beuys's most ambitious ecological project was a protest against the elimination of park, woodland and forest; his unique contribution to the seventh *documenta* in Kassel, Germany's biggest exhibition of visual art, entitled *7000 Eichen* [*7,000 Oaks*]. This was a tree-planting exercise conceived as sculpture on a gigantic scale, and, for Beuys, a natural and inevitable departure, which linked the world of the artist with the world of the ecologist,[12] a phenomenon which he described in an interview: 'a well ordered ideal of ecology ... can only stem from art ... Art is, then, a genuinely human medium for revolutionary change in the sense of completing the transformation from a sick world to a healthy one.[13]

'Art into life'

In 1974, Beuys established his own academy in which his aesthetic and anthroposophic plans could be realized; the Freie Internationale Hochschule für Kreativität und interdisziplinäre Forschung (FIH). The manifesto, which outlines the aims of the FIH, contains many of the ideas discussed above, and it also documents Beuys's concern with environmental issues. Most significantly, it states that the FIH was conceived as the basis upon which conditions could be established to develop and foster the creative potential that is present in all of us, but which is stifled in the modern world. To achieve this aim of breaking down the barriers between the arts and other spheres of activity, the institution was committed to interdisciplinary work; it was to become a laboratory in which art could be brought into social life and vice versa.

The immediate background for the project was Beuys's summary dismissal from his academic post following a protest against the *numerus clausus* system of controlling the admission of students to higher education. Beuys argued at the time that this act was evidence of the indifference of the state to aesthetic education. In his search for support and publicity for his new venture, Beuys turned to the writer Heinrich Böll (1917–85), who was also to become one of the high-profile supporters of the Green Party.[14] The two men had first been introduced in the 1960s, and they clearly had a productive exchange of ideas over many years. The original manifesto outlining the aims of the FIH does not acknowledge authorship, but when it was reproduced in the catalogue of a major exhibition on German art at the ICA, in London in 1974, entitled *Art into Society, Society into Art,* it carried both men's names.

The genuine extent of Böll's input, however, is questionable. Apart from helping to win media coverage, his main contribution would appear to be in one area of the syllabus, *Wörtlichkeitslehre*, a concept which Beuys directly attributed to Böll during a press conference accompanying the FIH's launch. Böll made himself available to teach this discipline, which he defines in terms of attempting to draw attention to the discourse of certain social groups, and to provide assistance in resolving semantic differences.[15] The concern with the precise meaning of words reflects Böll's acute awareness that linguistic conflict can initiate social conflict, a phenomenon which he expounded in numerous essays, and which plays an important role in some of his narrative works, such as the story *Die verlorene Ehre der Katharina Blum* [*The Lost Honour of Katharina Blum*] (1974), a satirical exposé of the gross manipulation and distortion of language by the tabloid press.

While some of Böll's own statements on the status of art in society can be employed to place Beuys's aesthetics in sharper relief, they also reveal that the two men had far less in common than their collaboration might suggest. Beuys's extended or expanded concept of art ('erweiterter Kunstbegriff') is based, as I have described, on a view of mankind which attaches prime importance to promoting the individual's creative capacities. It is based on an epistemology which sees all knowledge as deriving from art. Böll's views, on the other hand, are far less radical. Whilst Böll shares Beuys's tenet that art can widen the horizon of the beholder by revealing aspects of life which scientific discourse and other 'factual' accounts of reality ignore, in his epistemology, art does not have a monopoly of knowledge: the insights of art can merely complement those of science by exploring the 'gaps' which make up a realm beyond the façade of facticity, and which cannot be penetrated by scientific method. Böll was also careful to point to the dangers of rejecting reason: without such a balancing factor, he maintained, imagination can all too easily lose touch with concrete reality.[16]

Aesthetic education

Having discussed the main elements in Joseph Beuys's eco-aesthetics, I should now like to place them in a wider historical context, and identify certain ideas from the German tradition which exerted an influence on him. Clearly, Beuys's critique of our unrelentingly destructive way of organizing social and economic life has much in common with the political, sociological and ecological studies to which a number of the other contributions to this volume refer. Of undoubted importance, for example, is one of the seminal works of the Frankfurt School, Max Horkheimer and Theodor Adorno's *Dialektik der Aufklärung* [*Dialectic of Enlightenment*]. It would be erroneous, however, to assume that Beuys had at any time an orthodox left ideology. Indeed, he often distanced himself explicitly from materialist philosophy, which is hardly surprising, given that it was the materialist element in scientific method which he felt neglected mankind's spirituality, and which led him in the mid-1940s to regard art as his real vocation. A statement in an interview during one of his high-profile visits to the USA can be adduced as further evidence of his anti-materialism: 'I myself determine history . . . it is not history that determines me. Economic circumstances do not determine me. I determine them[17] . . . In the simplest terms I'm trying to reaffirm the concept of art and creativity in the face of Marxist doctrine.'[18]

For Beuys, creative capacity and self-determination are the only possible agents of real social change; an inner revolution is the necessary precondition for an outer revolution. Herein lies man's freedom, which is the freedom to create both his inner and his outer life. Beuys hoped that free, creative beings would see themselves as creative elements within an all-embracing social organism or sculpture. In so doing they would make a contribution to solving the environmental crisis by their developing a less reified relationship with nature. In other words, an individual is a microcosm of a universal macrocosm. This is an idea which, as several commentators have pointed out, is present in antiquity, in the humanism of the Renaissance, as well as in the holistic and natural philosophy of Goethe and the Romantics.[19] Moreover, Beuys's call for the '*Gesamtkunstwerk* of the Future Society' echoes the Romantic global concept of art, and what might be termed his 'will to art' picks up a thread which goes back to Friedrich Nietzsche.[20] However, the central importance of creativity as a means to overcome the alienation of man has its main antecedents in the philosophical idealism of the eighteenth and nineteenth centuries, particularly in the thought of the German poet and philosopher Friedrich Schiller.

Beuys's statement in an interview that 'Man is only truly alive when he realizes he is a creative, artistic being' is clearly reminiscent of Schiller's famous dictum in his *Briefe über die ästhetische Erziehung des Menschen* [*Letters on the Aesthetic Education of Man*]. Indeed, the questions Beuys raises demonstrate, not least, the abiding relevance of some of Schiller's deliberations. Like Beuys, Schiller's aesthetic position is preceded by critical reflections on society. In Schiller's case, of course, the point of departure for such criticism is the alienation of man as a result of the historical necessity of specialization and the division of labour. Because the process of the division of labour and the advantages it has brought to society as a whole are irreversible, another tool must be found to restore the lost totality, a totality which Schiller finds in its ideal form in the Greek city-state (*polis*). This tool is art. Schiller regards the 'Veredelung des Charakters' [ennoblement of the character] of the individual in the realm of the aesthetic to be the agent of the 'Verbesserung im Politischen' [improvement in the political]. One major difference from Beuys's views, however, can be identified: aesthetic education in the manner Schiller prescribes requires the recipients to be capable of understanding and appreciating works of art, and is thus restricted to the subculture of an educated élite.[21]

For Beuys, of course, with his belief that every man is an artist, such a bourgeois view of 'high art' is anathema.

The social mission of art can also be found in Expressionist thought, and is exemplified in works such as those by the sculptor Wilhelm Lehmbruck which Beuys so admired.[22] It is in this context that a link can be made to a figure whom I consider to be the key inspiration for Beuys's expanded concept of art. It was through his interest in Lehmbruck that Beuys became acquainted with the teachings of Rudolf Steiner (1861–1925), the Austrian social philosopher and founder of anthroposophy.[23] Steiner's aim was to integrate the psychological and the practical dimensions of life into an educational, ecological and therapeutic basis for spiritual and physical development. One document is particularly worthy of attention: Steiner's *Aufruf an das deutsche Volk und die Kulturwelt* [*Appeal to the German People and the Cultural World*] (1919), in which he sets out his theories for the regeneration of society, following the mass carnage of the First World War. (Wilhelm Lehmbruck was, along with Hermann Hesse, one of the signatories and committee members of this organization.)

The key to Steiner's thinking is his demand for a 'threefold social order'; (the 'Dreigliederung des sozialen Organismus'). Steiner believed that a humane social life could only be restored if the state, the economy and intellectual life were kept separate. The principles to build a completely new foundation for the social organism were essentially those of the French Revolution: equality before the law, fraternity, and liberty. The economy was to be regulated not by the state but by brotherly co-operation between all participants in the economic process; producers, distributors and consumers. Finally, art, science, religion and education were to be governed by the principle of liberty. Beuys's all-embracing theory of 'social sculpture' is clearly influenced by these views, which are echoed in many of his theoretical statements, not least his own *Aufruf zur Alternative*. Like Steiner, Beuys also demands a change in the function of money, so that production will be determined by the needs of the consumer rather than by profit. Little wonder, then, that one of Beuys's earliest political groups, the Organization for Direct Democracy had Steiner's concept of the 'threefold social order' in its full title, and it is of particular significance, in this context, that Beuys should acknowledge the seminal importance of Steiner's teachings for his own anthroposophic and aesthetic credo in a public lecture given shortly before his death in 1985.[24]

Conclusion: '... to be a teacher is my greatest work of art'[25]

The last years of Beuys's life witnessed a growing disenchantment with the Green Party. Although Beuys's name headed the list of Green candidates in North Rhine-Westphalia following the constitution of the Greens as a federal party in January 1980, he was unceremoniously removed, amid a welter of publicity, shortly before the elections. The reasons for this move are, I believe, quite complex. Given that Beuys's eco-aesthetics were, as I hope to have demonstrated, an eclectic cocktail of ideas, it is not hard to imagine that his anti-materialism and his anti-rationalism, together with a bold mixture of pagan and Christian mysticism, found little favour on the wing of the party that contained residual Marxist elements of the Student Protest Movement. Another reason lies perhaps in the fact that Beuys would never be an easy partner for a party which was preparing to assume a completely normal place within the Federal Republic's political structure, as one of the co-founders of the FIH has suggested.[26] Certainly Beuys lost no time in attacking the Greens' 'capacity for politics' after 1980 as a renunciation of '... all potential for forward-looking ideas'.[27] Whilst he was undoubtedly an important mediator of key ideas which have filtered into modern ecological thought, therefore, his association with the party was ultimately a frustrating one.

The rather unhappy end to Beuys's formal links with the Greens raises the question of the extent to which his influence, particularly strong when the movement was in its infancy, has endured. His view that aesthetics is integrally connected to the problems of society, and that an expanded concept of art is the medium for restoring the organic totality of the individual, society and nature, is a common thread in many of the arguments advanced in green debates on culture, and is likely to remain his most profound contribution.[28] There are, however, also other areas in which his voice can still be heard, not least in the activities of the three state-funded, and so-called 'grünnahe' political foundations [foundations sympathetic to the green movement]. These were set up in 1987, when the erstwhile Green MP and lawyer Otto Schily won a test case in the courts to uphold the right, enshrined in the German constitution, of every 'dauerhafte gesellschaftliche Strömung' [lasting social trend] to be represented in this manner. One such institution, for example, the Heinrich Böll Foundation, was managed for several years by the influential ideologue Lukas Beckmann, who is only one of many of Beuys's pupils to assume positions of power within the party. Beckmann has written glowingly of the importance of his mentor to the development of the Greens,[29] and there can be little doubt that he and other Beuys acolytes within the Green hierarchy continue to regard themselves as the custodians of a still vital legacy.

Notes

[1.] Johannes Stüttgen, *Der erweiterte Kunstbegriff und Joseph Beuys' Idee der Stiftung,* (Cologne, 1990), 34.

[2.] Irit Rogoff, 'Representations of politics: critics, pessimists, radicals', in *German Art in the Twentieth Century* (London, 1985), 129.

[3.] Heiner Stachelhaus, *Joseph Beuys* (Munich, 1989), 152.

[4.] For an English translation of the text see Caroline Tisdall, *Joseph Beuys* (London, 1979), 283–4.

[5.] Robert Hughes, *The Shock of the New: Art and the Century of Change,* updated and enlarged edn. (London, 1991), 400.

[6.] Joseph Beuys, 'Interview with Kate Horsfield', in Carin Kuoni (ed.), *Energy Plan for the Western Man: Joseph Beuys in America. Writings by and Interviews with the Artist* (New York, 1990), 64.

[7.] The dead-end into which 'bourgeois' art and culture had careered was to become the subject of many of his works in the early 1960s. On a number of occasions it is the concert grand piano which Beuys deploys as a monolithic symbol of bourgeois art's social redundancy in such famous works as *Infiltration homogen für Konzertflügel* [*Homogeneous Infiltration for Piano*].

[8.] Joseph Beuys, 'I am Searching for Field Character', in Kuoni (ed.), *Energy Plan for the Western Man,* 21 (Beuys's own upper case).

[9.] See Tisdall, *Joseph Beuys,* 190.

[10.] Shamanism was a form of religion, practised primarily in Siberia but also in Greenland, in which the world is governed by good and evil spirits which can be appeased by the intervention of the shaman, a priest or sorcerer. Beuys's interest in shamanism is dealt with by Tisdall, *Joseph Beuys,* 22–3.

11. Hughes, *The Shock of the New*, 405.

12. Hiltrud Oman, *Die Kunst auf dem Weg zum Leben: Joseph Beuys. Mit einem Essay von Lukas Beckmann* (Berlin, 1988), 158.

13. Joseph Beuys, 'Time's thermic machine; a public dialogue, Bonn 1982', in Kuoni (ed.), *Energy Plan for the Western Man*, 99.

14. It is important to note, however, that it was with the Greens' stance on the disarmament issue that he had the greatest sympathies.

15. See *Der Tagesspiegel*, 23 February 1974, which quotes Beuys as saying 'Heinrich, that's your term, you can explain it.'

16. Heinrich Böll, 'Versuch über die Vernunft der Poesie', in Bernd Balzer (ed.) *Heinrich Böll. Essayistische Schriften und Reden III* (Cologne,1978), 34–50.

17. Joseph Beuys, 'Interview with Willoughby Sharp', in Kuoni (ed.), *Energy Plan for the Western Man*, 86.

18. Ibid., 90.

19. See Peter-Klaus Schuster, 'In search of paradise lost. Runge – Marc – Beuys', in Keith Hartley et al., *The Romantic Spirit in German Art 1790–1990*; and 'Man and his own creator; Dürer and Beuys – or the affirmation of creativity', in *In Memoriam Joseph Beuys. Obituaries, Essays, Speeches*, tr. Timothy Nevill (Bonn, 1986), 17–26.

20. A long-standing preoccupation with Nietzsche emerges in a number of interviews and is touched upon by one of his biographers. Beuys made a pilgrimage as a young man to the Nietzsche archives in Weimar, and in 1978, he produced a work entitled *Solar Eclipse and Corona*, which is dedicated to the philosopher.

21. Rolf Grimminger, 'Die ästhetische Versöhnung. Ideologiekritische Aspekte zum Autonomiebegriff am Beispiel Schillers', in Jürgen Bolten (ed.), *Schillers Briefe über die ästhetische Erziehung des Menschen* (Frankfurt am Main, 1984), 161–84 (here, 179).

22. Beuys recalls first becoming interested in art as a seventeen-year-old when he was impressed by a photograph of a sculpture by Wilhelm Lehmbruck, and it was from this point on that he had an inkling of art's potential. Beuys, 'Interview with Kate Horsfield', 64.

23. For the impact of Steiner's thought on Beuys see his interview with Louwrien Wijers in Kuoni (ed.), *Energy Plan for the Western Man*, 215–58. The relevant sections are on pp.255–9. For a comprehensive analysis of Steiner's thought see Gerhard Hahn, *Die Freiheit der Philosophie. Eine Fundamentalkritik der Anthroposophie* (Göttingen, 1995).

24. Joseph Beuys, 'Thanks to Wilhelm Lehmbruck', in *In Memoriam Joseph Beuys*, 57–61.

25. Joseph Beuys, 'Interview with Willoughby Sharp', in Kuoni (ed.), *Energy Plan for the Western Man*, 85.

26. Klaus Staeck, 'Democracy is fun', in *In Memoriam Joseph Beuys*, 12.

[27.] Joseph Beuys, 'Talking about one's own country: Germany', in *In Memoriam Joseph Beuys*, 47.

[28.] For a full discussion see Stephen Brockmann, 'The green battlefield: aesthetics and "Kulturpolitik" in the current debate', *German Life and Letters*, 43, 3 (1990), 280–9.

[29.] Lukas Beckmann, 'Joseph Beuys – begreifen, nicht verdrängen', in Oman, *Die Kunst auf dem Weg zum Leben*, 179–95.

15 • Eco-drama in the *Naturtheater*: The Work of Martin Schleker

Alison Phipps

An Introduction to *Naturtheater*

Naturtheater is a form of open-air community theatre. The term *Naturtheater* is a regional one, confined to the south-west of Germany, with occasional Bavarian or Rhineland exceptions. The usual German term for *Naturtheater* is *Freilichttheater* or *Freilichtbühne*. These terms come close to the sense embodied in the English translation of 'open-air' or 'outdoor' theatre. However, a translation of the term *Naturtheater* becomes problematic when an attempt is made to transmit the sense of nature and of local colour embodied in the German term.[1]

Naturtheater, to a Swabian, is an amateur performance which takes place in settings perceived to be natural or picturesque. These settings may include medieval market squares, castle walls, ruins or cathedral courtyards; or they may be set on the outskirts of a town in an abandoned quarry, a gorge, by a stream or in woodland. The *Naturtheater* settings are also invariably located in public rather than private space. The amateur status of the performers makes *Naturtheater* productions community events, drawing on the talents and goodwill of the local population. The local base of the *Naturtheater* is reflected both in the style of production and in a marked preference for performing in dialect and alluding to local places and events: local celebrities and local politics may be identified. The plays performed in the *Naturtheater* vary considerably; historical plays, popular plays or *Volksstücke*, musicals and the so-called classics, such as Schiller's *William Tell,* all occur in the *Naturtheater* repertoire, invariably delivered in Swabian dialect. Some theatres have also had plays written for them by local dramatists, productions which reflect local history and the concerns of local people. Local dramatists who have written for the *Naturtheater* in recent years include Erwin

Leisinger, Paul Wanner and, in particular, Martin Schleker.[2]

Popular theatre has often been dismissed as trivial or as a less than worthy focus for scholarly attention. However, the postmodern paradigm has opened up the study of literature, culture and society, and made the privileged position previously assigned to the literary canon seem untenable.[3] In particular, *Naturtheater* as a phenomenon of popular culture affords valuable insights into the way in which green thought has permeated German consciousness and grown in popular acceptance. The *Naturtheater* draw audiences from across the region they serve, and their popularity makes them an important social and cultural phenomenon. Audience members tend to be largely lower-middle-class in terms of their broad social status, but all social groups and all age groups are represented. Notably, very few of those who regularly attend *Naturtheater* performances attend other styles of professional theatre productions. It is the receptiveness of audiences to this style of theatre which makes *Naturtheater* an interesting focus for a discussion of eco-consciousness in German culture, and especially popular German culture. The community-based and locally resourced character of the *Naturtheater* makes them ideally placed not just to communicate environmental concerns but to put into practice a more sustainable lifestyle. Not all the *Naturtheater* in the south-west region of Germany, which formed the focus of this research, provided a locus for the active communication of environmental concerns, in spite of their often eco-friendly nature. However, one in particular, the Hayingen *Naturtheater*, may be regarded both as a vehicle for the communication of eco-consciousness through drama and as a practical example of a community akin to the sustainable models advocated by the green movement. The reputation of the Hayingen *Naturtheater* for eco-drama and alternative styles of popular theatre has developed since the advent of their resident dramatist, director and professional actor, Martin Schleker, in 1976.

A brief history of the Hayingen *Naturtheater*

The first *Naturtheater* production in Hayingen took place during the summer of 1949. The play performed was *The Organ Makers*,[4] and was written and produced by Martin Schleker senior. The site chosen for the performance was the natural limestone gorge just outside the village, in the Lautertal. Between 1949 and 1975 Martin Schleker senior established the Hayingen *Naturtheater* tradition in which performances became an annual event, attracting average audiences of around 500 per performance in the early years. Numbers grew to 1,000 as the tradition of performing became established. Every year saw the production of a new play and when, in 1975, Martin Schleker junior took over from his father in the theatre, the tradition of annual innovation was maintained until 1995. Today, annual audience figures regularly total between 25,000 and 30,000 and the *Naturtheater* has come to function as a small family business.

Schleker family involvement in the *Naturtheater*

The Schleker family involvement in the Hayingen *Naturtheater* is as much of a tradition as the annual performance of a new play. The importance of family involvement and the loyalty of the Schleker family both to the *Naturtheater* and to Hayingen is apparent from the cast lists. In the 1994 production Christel Schleker designed the elaborate set, while Martin Schleker with the two eldest daughters took three of the main parts. Although the Schleker family dominate the productions, many other members of the local community have taken demanding parts in the seasonal performances. To view the *Naturtheater* as purely a Schleker family undertaking would be to misrepresent both the diversity of people that make up the Hayingen *Naturtheater* and also the vision the Schleker family have of the theatre as a creative focus for a rural population in which issues affecting the lives of those in the *Naturtheater*, in the local villages and in the wider audiences, may be questioned and explored.

Martin Schleker: life and works

Although the Hayingen *Naturtheater* is a communal undertaking, the vision for its role as a provider of popular, political theatre which at once entertains and enlightens has developed under the direction of Martin Schleker junior.[5] Schleker's father had written plays celebrating local legends and local figures, yet which were steeped in the Nazi abuse of local and peasant art and fed by a desire to bring colour and life back to broken communities. By the 1960s and 1970s the folk ideologies of the Nazi past were being questioned, and Schleker junior returned to Hayingen from his acting training with a perspective on the function of art and theatre different from that of his father. Consequently, the transition of the *Naturtheater* from father to son was not a smooth one, either for the family or for the community, and there was much initial scepticism about Schleker junior's left-wing agenda for the *Naturtheater*. However, Schleker's new style of *Naturtheater* drama was highly successful, drawing large audiences and acclaim both from the press and from the regional government. Since 1975 Martin Schleker has won a regional drama prize[6] four times and established a regional reputation for his work in radio and television as well as in the Hayingen *Naturtheater*.

Martin Schleker junior has continued some of the work of his father, both in creating a new production every year and in his use of history, particularly local history, as the basis for his plays. However, Schleker's intention, in using historical material, differs from his father's. He seeks not merely to entertain, but also to demonstrate parallels between past and present, in particular with respect to the lives of ordinary people. In the 1970s and early 1980s his interpretation of history was a product of his left-wing convictions: the peasants, who had been exploited by the aristocracy and by the wealthy Catholic church in the past, were still being exploited by others in positions of power today. In his work he differs from authors of contemporary popular plays (*Volksstücke*) such as Kroetz or Sperr in that he writes for a local working- or lower-middle-class audience, rather than for the educated classes,[7] and unlike many contemporary German dramatists, does not see humour as misplaced where human drama is played out in the theatre. On the contrary, Schleker maintains that humour, a roughness of both language and acting, and a sense of irony and of fun, similar to those used in the cabaret tradition, may act as more effective vehicles of criticism than a theatre that alienates the audiences it seeks to change.[8] In this respect he succeeds in avoiding the nostalgia and the romantic associations other *Naturtheater* in the region seek to cultivate. Schleker's adaptations of Wilhelm Hauff's novel *Lichtenstein*[9] in 1981 enabled him to confront his audiences with romantic clichés and to demonstrate that the majesty of Duke Ulrich of Lichtenstein was only bought at great cost to the exploited peasant community. In this sense Schleker works in the tradition of Büchner and Brecht, for his plays have consistently challenged the myths of the past and the injustices of the present. In recent years, however, Schleker's work has developed a concern with wider social issues. His socialist stance now incorporates feminist and green agendas, and his use of stereotypical characterization and alienation techniques has lessened in favour of an exploration of the complex emotions and motivations of his characters. His recent adaptation of Shakespeare's *A Midsummer Night's Dream* demonstrated his ability to incorporate both the epic and the dramatic styles of theatre,[10] allowing both psychological tension and a political message.

Schleker's development as a dramatist has not always been welcomed by critics from the local press. For many years Schleker was seen as a champion of the left-wing cause, and his new-found holistic pluralism and the deepening emotional content of his plays have left some former allies feeling betrayed:

> There were no rebel angels. Everyone agrees that when you go to Hayingen you know what to expect . . . But this year the supper was a little more meagre than we're used to . . . Sparks of genius were rare, and, except for a 'chains of light are all well and good, but those who have nothing take their anger out on the even less fortunate' thrown in at the end, we waited in vain for political references.[11]

This extract, from a respected left-wing broadsheet newspaper, reveals both the limited nature of local journalism and the restrictions of ideology in criticism. The function of this production, and equally of other productions in other *Naturtheater* in the region at a cultural rather than an intellectual or literary level is completely overlooked. The play in question is Schleker's adaptation of Sailer's baroque play *The Swabian Creation Story*,[12] in which issues of patriarchy, religious domination, unfair distribution of wealth and the breakdown in relationships between husband and wife are explored in depth. It explores social constructs such as the masculinity of God along gender lines, and through its set design it keeps both the stark contrast between the First and the Third World and the endangered species of the rain forests visible to the audience.

Eco-texts[13]

The environmental issues which have been a feature of Schleker's work for over ten years complement his socialist beliefs rather than threaten his original political persuasions. For Schleker, the social and the ecological are interdependent, and in several plays, particularly in the 1990s, Schleker has drawn clear parallels between the destruction of society through greed on the one hand and the ecological crisis on the other. Clearly, Schleker's pacifist, anti-nuclear stance, a feature of his play *Das Dorf auf der Grenze* [*The Border Village*], has informed his growing concern for environmental issues. In 1986 the Chernobyl disaster required an environmental as well as a socialist reaction, and his play *Rulaman*[14] contained a reference not just to the destructive effects of war and the danger of the arms race, but also to the dangers of environmental destruction. In an interview Schleker maintained that he had changed the play, following the Chernobyl disaster, from bearing a purely pacifist message and reflecting the growth in the peace movement during the stationing of US Cruise and Pershing missiles in Germany.[15] He had originally intended to have an unrealistically happy ending to the play in which the warring factions would turn their swords into trumpets, rather than ploughshares. However, instead he ended with a song, sung by the soldiers and following in the epic tradition, injecting a sense of tragedy into an otherwise comic play and refusing to give his audiences the same false sense of security as politicians had attempted to do in the wake of the Chernobyl disaster:

> Es war ein Land, voll Blumen, Milch und Honig
> Und Fleisch und Butter, berghoch.
> Da kam der Tag, an dem wir schneiden mußten
> und alles, alles ließen wir zurück:
> die sanften Hügel, die grünen Wälder
> die stillen Gassen, das traute Glück.
> Es gab noch immer Blumen, Milch und Honig
> und frischen, grünen Kopfsalat – bloß:
> essen konnt man ihn nicht.

[There once was a land, full of flowers, milk and honey/ And meat and butter, piled high./ Then came the day we had to make a break/ And we left everything behind:/ the gentle hills, the green woods/ the still alleyways, the certain happiness./ There still were flowers, milk and honey/ and fresh green salad – only:/ we couldn't eat it.][16]

Instead of celebrating their peace the newly reconciled peoples disappear underground, into one of the many limestone caves, a symbolic feature of the local countryside, vowing not to emerge until several thousand years have passed and the danger of radiation has gone.

Rulaman represents a turning point in Martin Schleker's work, for he developed the parallels between the exploitation of the peasants in the past and the working classes of the present to show environmental exploitation as well. This change also appears in the programme notes Schleker makes for his audiences, explaining the motivations for his writings, detailing the genesis of the play and informing his audiences of developments in the Hayingen *Naturtheater*, thus demonstrating his belief in the *Naturtheater* as a community-based undertaking in partnership with audiences. In his description of the set for *Rulaman* and of the difficulties in creating a virgin landscape, as would have been appropriate for a play set in the stone age, Schleker refers to the number of fallen trees used in the set and shows an awareness of the concerns of his audience for the preservation of woodland and the problems caused by acid rain:

And the fallen trees on our stage are also 'natural'. We didn't fell them ourselves, they have simply died on us. Unfortunately this is true for more and more trees each year. The new wood is growing up in their place, but it needs time. Until then the audiences sometimes have the sun in their eyes.[17]

Schleker here shows how the quality of a *Naturtheater* experience can be affected by environmental pollution, and makes clear his position regarding the destruction of the German forests. In an interview in 1995, nine years later, Schleker's stance had developed further. He saw himself as a green voter, no longer a pure socialist, and saw one of the tasks of the Hayingen *Naturtheater* as the preservation and conservation of the *Naturtheater* site: 'The Tiefental is not an official conservation area but it is protected by us.' This was also borne out by the life-style of the *Naturtheater*, where wildlife was respected, much material was recycled and green issues were regularly raised by members of the casts. The yellow sacks, so much part of the Swabian institutionalized re-cycling culture, were used in the set design in 1993 as a visual reminder to the audience of the waste and recycling problem. Audience members are now encouraged to sort their interval rubbish into glass, paper and plastics. When the municipal council refused to send a lorry into the steep gorge to collect the rubbish, the cast volunteered to make sure that the sacks were transported out of the gorge.

Schleker's first real eco-drama came in 1991 with *Fast wie im Himmel* [*Almost as in Heaven*]. Green issues provide the context and the motivation for the play, unlike Schleker's earlier works where ecology seems more of an add-on. *Almost as in Heaven* focuses on local environmental issues, thus making abstract concerns concrete and accessible for the Hayingen audiences. In the introduction to the play, Schleker cites an extract from the regional press:

> The National Park of Berchtesgaden in the Bavarian Alps may use the accolade of 'Biosphere reserve' in future. Unesco has thus given the National Park, created in 1978 and surrounded by the Watzmann and the Königssee, the honour of standing alongside 300 other outstanding conservation areas.[18]

This news item is of no particular consequence to those involved in the Hayingen *Naturtheater*, or to the Hayingen audiences. However, Schleker makes it relevant, by introducing the idea that 'Number 301 on the Unesco list is the Swabian Alb', and suggest-ing that other economic plans could affect the future of the natural beauty and long agricultural tradition of the Alb: 'Will the Alb, with the Rhine–Neckar–Erms–Lauter–Danube canal, become a blos-soming industrial area with its centre in the international port of Indelshausen?'[19] To a non-Swabian, this question may have little personal resonance. But for those attending performances, from the Alb or directly from Hayingen and the surrounding villages, the idea of turning the Lauter, a small brook running through the famous gorge in which the *Naturtheater* is to be found, into a canal, or making the neighbouring farming hamlet of Indelshausen into an international port, both personalizes the problems of economic expansion for the environment, and allows Schleker to tackle envir-onmental issues through comedy. Throughout *Almost as in Heaven* the concept of progress at the expense of the environment and of local Swabian communities is ridiculed. Schleker offers utopian solutions to environmental problems, exposing the inequalities, injustices and often farcical nature of the environmental policies practised at local, national and global level. At the end of *Almost as in Heaven* the *deus ex machina* solution is one where ecological concerns override economic expansion. The gods Jupiter, Juno and Mercury announce to the assembled masses that their jobs and their environment are safe:

Jupiter: The Swabian Alb will become a conservation area and the inhabitants of the Alb will be state-employed conservationists.[20]

In spite of his roots and his commitment to the local community of Hayingen, Schleker has not limited his dramas to an exploration of local ecological issues. In 1994 *Der schwäbische Sommernachtstraum* [*The Swabian Midsummer Night's Dream*], Schleker's free adaptation of Shakespeare, took the global ecological crisis as its context. The feud between Oberon and Titania is used by Schleker as the catalyst, not just for the confusion between the lovers on earth[21] but also for the increasing threat of global ecological destruction. Oberon becomes a powerful yet disinterested despot able to watch human destiny and destroy the human race on a whim. A parallel is drawn between the Oberon figure and world leaders whose own self-interest and lack of action have created conditions of poverty and environmental destruction:

Oberon: I don't ask anything of that bunch of miserable human beings, they're going under anyway, like the whole world. One strike and everything will be ashes.

. . .

Titania: Then there'll be hell to pay. Nature will go mad. Humanity will go crazy. Earthquakes, floods. Sailing through the medieval quarter of Cologne by canoe. Misery and nerves. Famine and the bloodbath of war.[22]

The religious tone of the apocalyptic warnings in Schleker's play is deliberate. Schleker professes a radical sceptical Christian theology but sets himself up in opposition to the way he believes church leaders, particularly in the Roman Catholic church in the past, have manipulated their flock and failed to address pressing issues such as poverty, injustice and the ecological crisis. He is deliberately provocative towards a largely Catholic audience in this respect. In *The Swabian Midsummer Night's Dream*, for example, several members of the audience considered his use of the hymn *Nearer, my God, to thee* to be sacrilegious in the scene where Puck lights the fuse attached to a model of the globe. However, in spite of his radical stance and the criticism he receives from a minority of his audience members, Schleker maintains that the *Naturtheater* is a suitable forum for his own eco-politics and for his hope and belief in possible solutions. The audiences are sent away at the end of *The Swabian Midsummer Night's Dream* amused and relieved by Schleker's utopian ending,[23] yet with the words of his final speech impressing upon them the need to work for a better world, however much apathy and despair may prevail.

Conclusion

It is difficult to quantify Schleker's success in terms of his promotion of eco-consciousness. Certainly, as already demonstrated, his cast are particularly eco-conscious and actively environmental. The community focus of the *Naturtheater*, in which Schleker has a passionate belief, is also an important aspect of the overall eco-basis of Schleker's work. The sense of the importance of the *Naturtheater* community among the cast is equally passionate, although subliminal and rarely articulated. Members of the Hayingen community do not see their *Naturtheater* as a potential model for eco-life-styles, they see it as enhancing the quality of their lives and giving them a seasonal rhythm and purpose at a time when social cohesion in both rural and urban environments is in decline. Similarly with the audiences, it is hard to tell whether Schleker's eco-dramas and radical politics are the main appeal or whether it is the spectacle of live animals, rough Swabian humour, competent amateur actors, the locally composed music and the informal picnic atmosphere of the annual performances which bring thousands of visitors to Hayingen every year. It is certainly not simply the political message, a requirement, it would now seem, for the state-subsidized production in the main theatres in Germany, that draws audiences into the theatre in Germany as a whole.[24] In the last twenty years the audience numbers in the state theatres have dropped drastically and audiences have expressed their disenchantment with the constant politicization of their theatre.[25] In the independent theatre sector, which includes the *Naturtheater*, audience figures have increased, and qualitative interviews undertaken in the *Naturtheater* in particular have demonstrated an appreciation among the audiences for the entertainment and for the spectacle offered.[26] However, in Martin Schleker's Hayingen *Naturtheater* the political message, particularly in the eco-dramas discussed here, does seem to have found a resonance with the audiences. The audience figures for the years in which ecological issues have informed the plays have been higher than at other times, particularly in the case of *Rulaman* and the evening productions of *The Swabian Midsummer Night's Dream*.

Members of the *Naturtheater* cast in Hayingen are already living relatively eco-friendly lives in community, as a result of their commitment to a locally based, creative, communal activity, and do not require the same help or the same stimulus as those whose lives in community are more fragmented. Perhaps part of the success and popularity of Schleker's Hayingen project is that it shares its experience with its audiences and sends them away, not necessarily convinced by the politics they have heard in the dramas, but both curious and excited by the experience of watching ordinary people like themselves share an aspect of their lives in community.

Notes

[1.] For this reason I shall retain the German original in this chapter.

[2.] Erwin Leisinger was the local dramatist for the town of Hornberg and responsible for the creation of the *Naturtheater* in Hornberg after the Second World War with his play based on a rewriting of local legend, *Das Hornberger Schießen* [*The Hornberg Incident*]. Paul Wanner, a regional dramatist, wrote popular plays in dialect for Swabian audiences. His plays were particularly popular during the 1950s but their popularity has diminished in recent years and they currently appear dated.

[3.] Anthony Easthope and Kate McGowan, *A Critical and Cultural Theory Reader* (Buckingham, 1992).

[4.] *Die Orgelmacher* [*The Organ Makers*] was based upon the history of the building of the famous great baroque organ in the local monastery in Zwiefalten.

[5.] Perhaps the nearest British equivalent would be the theatre of John McGrath; see J. McGrath, *A Good Night Out: Popular Theatre: Audience, Class and Form* (London, 1989).

[6.] The *Förderpreis des Landes Baden-Württembergs*, a prize awarded to regional dramatists, on an annual basis, to encourage their work.

[7.] While terms such as 'working class' or 'middle class' may seem outdated in the 1990s, no other satisfactory terms are as yet as readily accepted.

[8.] Lothar Schmidt-Mühlisch, *Affentheater: Bühnenkrise ohne Ende?* (Frankfurt am Main, 1992), demonstrates this point well and shows that in the last twenty years German audiences, drawn from the middle classes, have abandoned state theatre productions in favour of independent and amateur theatre productions such as those offered by the *Naturtheater*.

[9.] Hauff was a member of the Swabian Romantic school.

[10.] Ulrich Staehle (ed.), *Theorie des Dramas* (Stuttgart, 1973), 68–98.

[11.] 'Aufstand der Engel fand nicht statt', *Schwäbisches Tagblatt*, 29 June 1993 (author unknown).

[12.] Sebastian Sailer, *Adams und Evens Erschaffung, und ihr Sündenfall* (Biberach, 1977), facsimile edition of original manuscript in the monastery at Obermarchtal, near Hayingen.

[13.] The texts of plays referred to are unpublished manuscripts in private possession or available from the *Naturtheater* in Hayingen.

[14.] Based on an adaptation of David Friedrich Weinland's Swabian best-seller.

[15.] 'Die Hayinger und "ihr" Martin Schleker', *Reutlinger Nachrichten*, 22 July 1986 (author unknown).

[16.] Schleker, *Rulaman*, 64.

[17.] From programme notes to *Rulaman*.

[18.] Extract from *Südwestpresse*, 28 March 1991, cited in Schleker, *Fast wie im Himmel* (Hayingen, 1991), 3.

[19.] Ibid.

[20.] Schleker, *Fast wie im Himmel*, scene 50.

[21.] *Hermia, Lysander, Helena and Demetrius*.

[22.] Schleker, *Der schwäbische Sommernachtstraum* (1994), 25–6.

[23.] Two boys urinate on the lighted fuse attached to the globe, thus averting the apocalypse.

[24.] See Schmidt-Mühlisch, *Affentheater*.

[25.] Ibid.

[26.] Statistics obtained from the *Verband Deutscher Freilichtbühnen*, Hamm.

16 • Tatort Umwelt: Crime Fiction as Ecological Agitprop[1]

Birgitta Schüller

> Popular fiction is a barometer of social attitudes. Of course all fiction betrays the fears and wishes of its writer. But because crime fiction focuses so strongly on issues of justice and society, characters have to be exaggerated to make social points. People behave in more stereotypical ways than they might in mimetic fiction, because their actions have to depict concepts of power, law and justice.
>
> Sara Paretsky[2]

The German *Krimi*, the detective novel, responds to currents of social change faster and better than other literary genres.[3] It reflects the reality of the Federal Republic with a remarkable density of detail, making it a mine of information about the country, as well as a faithful mirror of social processes and of the cultural *Zeitgeist*.[4] With novels featuring Turkish detectives pursuing killers of asylum-seekers, novels set in cities like Cologne and Düsseldorf, regional *Krimis*, or Pieke Biermann's novels about the Berlin gay scene (which have won several awards),[5] crime fiction in German is enjoying a renaissance, unashamed of its reliance on English-language models.[6]

This chapter concerns environmental issues – environmental protection and environmental crime – in detective fiction. What differences can be observed between the so-called 'first generation' of authors such as Lydia Thews, Horst Bosetzky (using the pen-name '-ky') and Michael Molsner, and the 'second generation', represented here by Viola Schatten and Jakob Arjouni? The first generation aspired to make crime fiction a vehicle of social enlightenment, using the device of suspense to 'sensitize the reader to social problems'.[7] But can this ambition, specifically in relation to 'green' issues, be squared with the demands of mystery and suspense writing? How does the motif of 'the environment' function in these novels, either as a *Tatort* [scene of a crime], or in the life of the fictional characters? Is social reality reflected accurately when the detective is a politically correct hero(ine), who rides a bicycle, does not smoke, tends a herb garden and eats nothing but organic grains and pulses?[8] Is it possible to speak of a distinct subgenre of environmental crime fiction?

The surprise success of the early 1990s was Viola Schatten, who has so far published three novels featuring days of the week in the title.[9] Critics have vied to discover who is behind the pseudonym, naming authors such as Ulla Hahn, Eva Demski, Cora Stephan and Eva Heller, and even comparing the novels with Ricarda Huch's classic *Der Fall Deruga* [*The Deruga Case*].[10] Schatten's detective is an impoverished Bavarian noblewoman, Ruth Maria von Kadell, who has studied psychology and philosophy, teaches ju-jitsu, and is described by a male critic as 'clever, beautiful, fast, skilled in self-defence and erotically mobile'.[11] The *Süddeutsche Zeitung* is cited on the back of the 'Tuesday' book, declaring Schatten to be writing '*the* crime novels of postmodernism'. Left liberal fashion dictates that a mantle of environmental concern be worn over the trenchcoat borrowed from Hammett and Chandler, and while none of Schatten's books feature environmental crime, they all include comments on 'green' issues alongside descriptions of gourmet meals and designer clothes.

In Schatten's 'Tuesday' novel, von Kadell investigates both a theft of antiques in Frankfurt am Main and a spectacular murder, uncovering conspiracy and corruption among local and national politicians in both main parties. The Frankfurt book fair provides local colour, and there are suggestions of a *roman à clef*. It is revealed that the detective, as a student, took part in the demonstrations against the construction of the new west runway at Frankfurt airport, and that she now votes green, 'although they [Die Grünen] have gone to sleep and a renewal is long overdue' (*DN*, 80). She and her friends share a political stance described as 'clearly to the left, and that in European terms' (*MS*, 42). Politically correct consciousness is indicated on the first page of the 'Tuesday' novel: 'This year the start of October was fresh and chilly, despite the greenhouse effect and climatic catastrophe' (*DN*, 5). 'Miss Marlowe' from 'Mainhattan' (*DN*, 21, 62) may be the first bicycling detective: in this way she helps to save the environment, even though she is less combative than some other women cyclists in feminist crime fiction, who sabotage or destroy cars parked on cycle paths.[12] Kadell lives in a shared flat

(*Wohngemeinschaft*), and her daily routine includes sorting refuse into different bins for recycling (*DN*, 97). Moreover, the culture of contemporary environmental consciousness provides a stock of images and metaphors: 'Uli turned as pale as a sheet of recycled paper' (*DN*, 57); 'like a lousy biotope for threatened species, this pub' (*DN*, 71).[13]

Schatten's novels are far from being ground-breaking postmodern fictions. Their style is 'crude and superficial, a compound of clichés and quotations, [and] the plots are thin'.[14] They are unlikely to sell far beyond Frankfurt, where the inhabitants can enjoy recognizing local names and places in a caricatural guessing-game. The environment as theme has become one among other fashionable ingredients in these slick but undemanding books; the author and the heroine use the vocabulary of environmentalism in the same non-committal way as references to modern films and literature. But this kind of incidental integration of environmental consciousness does, after all, accurately reflect the realities of contemporary German everyday life, where time must be set aside for sorting household refuse, or for checking for the 'green dot' (*der grüne Punkt*) when shopping, and where words like 'biotope' have entered the vernacular.

The most interesting contemporary writer to deal with the environment as a scene of crime is Jakob Arjouni, in the second part of his 'Kemal Kayankaya' trilogy, *Mehr Bier* [*More Beer*] (1987).[15] Greeted as the 'best newcomer in German crime fiction',[16] and generally renowned as a Turkish-German *Wunderkind*, Arjouni has, so far, published four novels and several plays. He is 'the product of a liaison between two top-quality artists (the son of the publisher Ursula Bothe and the dramatist Hans Günther Michelsen)',[17] and it seems that the promotional legend of mixed Turkish and German parentage may have been launched by his Swiss publisher, Diogenes. The anti-hero of his three crime novels, a 'black-haired brother of Philip Marlowe'[18] and 'the most unusual detective ever to be at home in a German crime novel',[19] is the adopted child of two German schoolteachers, and speaks not a word of Turkish. Arjouni's local Frankfurt colour, occasional use of dialect, powerful plots and convincing characterization of the hero make an effective mix. The detective is no 'Robin Hood from Istanbul' (*EM*, 91), but a genuinely contemporary, complex figure.

The plot of *Mehr Bier* concerns environmental crime in the

chemicals/pharmaceuticals industry, and involves an intricate web of political intrigue. The head of a chemicals company has been killed by a mysterious bomb, while his firm faces a lengthy and expensive trial for injuries to children caused by a spillage of dangerous chemicals into a lake used for bathing. Another firm's attempt to open a new factory faces protests, demonstrations and occupations of the works by nearby residents, and rumours are abounding about exports of poison gas to Iraq. The murder is a boon to shareholders, including the mayor of Frankfurt, who is also a legal adviser to the chemicals industry, as public opinion quickly sides against the protesters: 'You can imagine how many people were happy that Böllig was bumped off by the Greens. Not because they were worried about the competition – his firm was totally insignificant – but because it meant that the chemicals industry had the martyr they needed' (*MB*, 49). A headline quoted in the prologue reads: 'After the Red Army Faction, now it's the Green one' (*MB*, 6). Prologue and epilogue are collages of newspaper headlines and reports about the murder, the revelation of the real guilty parties, and the minimal sanctions they suffer.

Four members of an organization called the Ecological Front are charged with the murder, and Kayankaya is asked to investigate by a left-liberal journalist and by their lawyer, Anastas, who is described as 'a kind of mixture of Gandhi and cottage in France. Gives his friends presents of wine, or Wallraff. I suspect he's in favour of free elections in South Africa' (*MB*, 81). Kayankaya has one week in which to find a police informer in order to solve the case. But he does not identify with his employers: 'If it makes you feel better, I've nothing against knitting your own socks, free-range chickens, women discussing things, or unsprayed raisins, and I've never liked wearing fur anyway. But don't ask me when the next collection of old newspaper takes place' (*MB*, 22). The detective's investigations in the fictional small town of Doddelbach reveal the story-teller's biting social criticism. For Arjouni as for Schatten, the German middle-class, foreigner-free garden suburb or small town is just a façade of decency concealing latent fascism. Allotments, picket fences, health sandals and the sorting of household refuse become murderously threatening symbols. Whereas Schatten describes the sorting of refuse in terms of harmlessly indulgent self-ironization, Arjouni associates a campaign to persuade citizens to sort their rubbish into orderly bins with a petition on behalf of the far-right

Republikaner party. Both detectives see the notorious self-ascribed national trait of *Gründlichkeit* [humourless thoroughness] at work in manifestations of popular German environmental consciousness, but Kayankaya sees behind the mask of the self-appointed concierge, the xenophobic and order-obsessed grocer who fumes:

> This empty packet of cigarettes, I found it on the stairs this morning! Because *I* sweep my stairs! Do you understand?! I sweep my stairs! Here in Germany people sweep the stairs outside their doors! It's not how it is in the Balkans, and you'll have to get used to it. Or you'd better go back there! You're terrorizing the whole house with your muck . . . yes, the whole house! (*MB*, 24)

Neither Die Grünen nor committed cyclists escape Arjouni's satire. In the third book of the trilogy, an ineffectual solidarity committee is fortunate in managing to prevent the detective from being deported:

> The whole of the SPD is in a studio, recording an election song, and the person in charge from *Die Grünen* is having a baby, or his wife is. His deputy has no car, but is coming as quickly as possible. That leaves the Multicultural Office. The phone was answered by a cleaning lady who hardly speaks any German, but as far as I could gather, they're all out at the opening of an exhibition about castanets. (*EM*, 135)

When his search for the security agent proves fruitless, Kayankaya suffers a nightmarish vision which seems to parody Grass's *Die Rättin* [*The Rat*][20] – 'A giant rat wearing underpants was sitting on the edge of the bed making a passionate speech about the German forests' (*EM*, 70) – before the police arrive, breaking down the door, beating him up and arresting him. Eventually a small-time crook and drug dealer helps Kayankaya break into police headquarters and steal a file containing the name of the informer. The members of the Ecological Front are finally convicted only of criminal damage, while questions about the political background of the murder 'remained unanswered' (*EM*, 171).

Arjouni uses contemporary environmental themes to paint a desolate picture of German reality. Schmidt notes that the novel is: '. . . half social tragedy, half environmental scandal that the German authorities, police and politicians cover up rather than investigate. The web of corruption is impenetrable, and Kayankaya gets caught up in it, too'.[21] In the prologue and epilogue, fictive newspaper reports ironically portray the career of the mayor of Frankfurt, from 'chair of the UN's environmental safety council' (*MB*, 7) to German presidential candidate (*MB*, 170). Who murdered or assaulted whom, to what extent corrupt policemen or eco-terrorists are guilty, is deliberately left unclear. There is no final monologue of enlightenment *à la* Agatha Christie; instead, the many layers of guilt and corruption lead the detective to despair.[22]

Arjouni and Schatten represent the second generation of 'new' German crime writers, while a selection of 'eco-crime' stories by the so-called first generation has been anthologized under the title *Tatort Umwelt* [*Crime-Scene: The Environment*].[23] It includes stories by -ky, Michael Molsner and Lydia Thews. The first of these writers has described crime fiction as a 'contribution to social analysis and critique',[24] while a chapter on his work in Jochen Schmidt's study of the genre is headed 'Die Misere des Soziokrimis' ['The poverty of socio-critical crime fiction'].[25] This judgement is confirmed by a reading of -ky's story in this anthology, 'Management by Nessoshemd'[26] (English in the original; *Hemd* means 'shirt'). It is set in Brasse, -ky's fictional north German town, and it depends on intertextual references to previous stories/cases, as do many of -ky's stories. The protagonist-detective is the police officer Reiner Reinsch (*rein* meaning 'pure, clean') who works for a special department, 'K 9: environmental and water protection' (MN, 9). Reinsch is an eco-chondriac, and his hygiene mania and fanatical concern for the well-being of the environment are described and explained at length in sociological and psychological terms. He suffers from various allergies and from dermatitis, while his two sons are asthmatics with various other respiratory and skin complaints. Luckily his wife is a former nurse:

> Her familiarity with children and doctors was especially useful to Reinsch, and after the birth of the second child she had had to concentrate wholly on her role as housewife and mother, and it was only now and then that she helped out in her friend Anke's organic vegetable shop. (MN, 23).

This is unintended parody from -ky, the self-proclaimed *Volksschriftsteller* [writer of and for the people],[27] whose 'poetics of everyday life',[28] coupled with lack of writing talent, leads to intolerable banality. He seeks to convince readers of the reality of his fictive (and stereotyped) locales and characters by piling up mimetic detail,[29] while at the same time using a demonstratively avant-garde narrative style which proclaims the fictional status of his text.[30]

The plot of 'Management by Nessoshemd' involves a local election in which a charismatic and popular SPD candidate stands against a CDU candidate who is a front man for the furniture manufacturing firm which dominates the town's economy. The former wants the derelict harbour area to be turned into an ecological nature park, while the latter proposes a technology park and a factory making plastic chairs and tables. Meanwhile a series of accidents and deaths occur, where the common factor is that the victims are wearing shirts made of artificial fibres produced by another local company: the 'death shirts' (MN, 47) produce a toxic-shock effect, by analogy with certain American-made tampons (MN, 41). Since the charismatic SPD candidate owns the company responsible for this chemicals scandal, he withdraws his candidacy, a political result which is the opposite of what Reinsch had hoped for. The chemist responsible for the formula is killed by one of his own shirts, his dying word being interpreted by Reinsch and his wife as 'Nessoshemd', referring to the poisoned shirt of Nessus which killed Heracles. It turns out that the shirt scandal was engineered by the CDU establishment to discredit the opposition: in -ky's work, 'the villains are always the "respectable people", members of the establishment.'[31] Reinsch's evidence is ignored or suppressed; the technology park is built in the face of local opposition; and it is finally revealed in flashback that Reinsch's eco-fanaticism led him to use the occasion of a celebrity football match to deliver a fatal kick to the CDU candidate's head. Reinsch now faces trial for murder.

This approach to the environmental theme condescends to the reader, and no suspense is felt as events unfold. As a writer, -ky sees himself situated 'somewhere between Jerry Cotton and James Joyce',[32] but his 'pseudo-lively style is supposed to overcome the unimaginative plotting and predictable solutions', and he fails in his ambition to use crime writing as a 'vehicle to repair our society'.[33]

Michael Molsner,[34] represented in *Tatort Umwelt* with 'Eine Stimme am Telefon'[35] ['A voice on the telephone'], also offers social criticism, but conveys it more subtly: it emerges in the thoughts and perceptions of his characters, in dialogue or stream of consciousness. This story involves his popular serial heroes, 'Euro-sleuths' Corinna Castrup and Markus Stauder, investigating murder and environmental crime in Petersried, an Alpine village. This tourism centre and natural park near Kempten is threatened by the plans of a multinational, Chemical United, to set up a plant in the vicinity. The opening pages introduce a large number of parallel plot strands, and only gradually is the reader able to make sense of it all, partly because the first-person narrator is a secondary charac-ter and the detectives do not immediately come into play. From a variety of perspectives, Molsner shows his mountain village becom-ing caught up in a net of guilt and crime. The plot and the motivation of the killer hardly seem realistic, but Molsner's local colour and characterizations are good.

A young investment adviser and leader of the Independents on the local council, Gotthold Justus Wolf, is entrapped by the chemicals firm with illegal share offers, blackmailed and then murdered when the corruption comes to light. Wolf is having an affair with Sylvia Voss, whose husband and daughter are environmental activists campaigning against the firm. Molsner describes them in caricatural terms reminiscent of Arjouni:

> Closing time for Meli Voss. She locked the door of her Third World Shop behind her and came over to me with her shy smile that said: I'd like to enjoy myself, but may I? An attractive girl, though unfortunately she hid her slim legs in baggy Peruvian cotton trousers and her breasts behind a Nicaraguan campesino shirt. At least she had let her black locks hang loose today, and they fell down over her Mexican man's jacket. I finished off the Senegalese peanuts she had sold me ten minutes ago and scrunched up the bag of recycled paper. (ST, 73)

Like -ky's officer Reinsch, Meli's father has developed an environ-mental mania which eventually brings destruction down upon the whole family. Meli, who is anorexic, is having an affair with the son of the *nouveau-riche* hotel owner, who is polluting the river with untreated sewage. Sylvia finally confesses to murder, her crime being attributed by the author to conflicts within the family caused by her husband's mania. But Molsner's story successfully shows how little remains of the mountain village idyll, even if in this case the multinational's plans are thwarted: 'Basically all were happy to be rid of the company; now other people could take on that mysterious giant' (ST, 141). In this short but complex story, Molsner is successful in maintaining a balance between enlightenment and entertainment, using Wolf, the liberal politician, as a mouthpiece to discuss the pros and cons of industrial development quite objectively.

Lydia Thews's story in *Tatort Umwelt*, 'Super-med'[36] (first published in 1986), is not her only contribution in this field. In 1984 she published an eco-thriller, *Störung der Totenruhe* [*Disturbing the Peace of the Dead*], in which a young journalist investigates illegal emissions from a chemical factory and uncovers a web of environmental and other crime in a small town.[37] Thews aims '. . . to write an exciting, entertaining story, though also to convey something' about 'the political and social problems of our time'.[38] 'Super-med' is half crime fiction, half science fiction. As with other writers discussed here, it is striking that her environmentalists are obsessive fanatics with criminal or terroristic leanings. In this story, they are prepared to murder innocents to achieve their ends, and references to the 'Green Army Faction' recall Arjouni's prologue. A group of environmentalist campaigners abducts Edgar Wilfing, described as 'the prototype of a 45-year-old senior manager in the car industry' (SM, 145), who is also a volunteer guinea pig for United Pharmaworks. The company is using him to test a new panacea and elixir of youth, 'Super-med', which has successfully cured his leukaemia. Thews convincingly portrays the opponents of this wonder drug and their tortuous ideological reasoning.

The leader of the group, Zulp, persuades his comrades to modify their demands by stages. After initially demanding that the experiments cease, he finally calls for the drug to be made freely available. Meanwhile one of the group, Fritzsche, breaks into the heavily guarded works and steals both a sample and a research report, which reveals that the body becomes dependent on the drug: 'If the public gets to hear about this, "Super-med" is finished, and you will have won' (SM, 202). This is entirely predictable, but there is a twist: Zulp turns out to be an agent planted by a rival drug company, and he now kills Fritzsche. Thews comments: 'My characters can accuse and bemoan but they cannot break free. They are always struggling – against circumstances and against themselves. And they always lose.'[39]

This seems true of all the writers and stories discussed here. It is striking that the chemicals industry provides the villain of virtually every piece, perhaps because of a cluster of major chemical spillages and scandals of various sorts in the mid- to late 1980s, when most of these texts were written and published. For most of the writers, the chemicals industry represents big capital. It may be that this industry is especially easily associated with crime and even murder because of a reputation for the blurring of borders (ethical and geographical) derived from complex cases involving the international movements of toxic wastes. But it is equally striking that the opponents of environmental crime, both as campaigners and as detectives, are not depicted as the 'true heroes' in pursuit of truth and enlightenment. Either they are naive and easily manipulated, or they are fanatics, murderers and eco-terrorists. And the multinational firms always win, at least in the long run.

Contemporary German crime fiction typically uses the crime-mystery framework as a vehicle for depicting and commenting on social issues. The mystery is commonly only partly solved, and little or nothing is said of its legal and social consequences (Arjouni's satirical epilogue is a partial exception). The detectives are broken heroes, who despair of their vocation, understand no more, or even less, than the reader, and frequently give up their job at the end of the case. Despite occasional partial successes, they are fundamentally powerless in the face of global economic interests and endemic corruption among the police and politicians.

As stated earlier, the environment as a theme figures in these fictions in two ways: as a factor in contemporary German everyday life and everyday consciousness (the hole in the ozone layer,[40] the green dot, etc.); and as the scene of a new, or at least newly significant kind of criminality. The first generation of German crime writers approached the theme in a spirit of caricature, while the second generation treats it more playfully and variously, but both share a pessimistic view of the possibility of arriving at the truth. The theme is only really successful when a case of murder (or at least manslaughter), as the essential focus of mystery, results from environmental crime. Crime fiction is not a suitable vehicle for ideas, unless they are fully integrated with the entertaining action.[41]

Many of the German 'eco-crime' publications discussed here are already out of print. In English, popular authors like Carl Hiaasen and Barbara Crossley seem to have assured the place of this type of writing as a distinct crime fiction subgenre, but it may be that it has only been a passing craze in Germany. Other themes are currently more fashionable, especially 'neo-nazism' and 'foreigners'.[42] But there is no doubt that as long as environmental concern and environmental crime are facts of German life, crime fiction will reflect those facts. '*Krimi* used to be synonymous with trash; now it stands for good entertainment', as Arjouni comments.[43]

Notes

[1] This chapter was translated from the German by Tom Cheesman. '*Tatort Umwelt*' means 'crime-scene: the environment'.

[2] Quoted in Evelyne Keitel, 'The woman's private eye', *Amerika Studien – American Studies*, 39, 2 (1994), 161–82, here 166.

[3] This chapter focuses on literary fiction, as does Peter Jordan in his article 'Carl Hiaasen's environmental thrillers: crime fiction in search of green peace', *Studies in Popular Culture*, 13, 1 (1990), 61–71. However, similar arguments could be made concerning film and television. Both television thrillers such as the *Tatort* series, and soap operas such as *Lindenstrasse*, tackle environmental issues. 'Deadly friendship', the *Tatort* broadcast on 23 May 1995, concerned the misuse of hormones in farming, contaminated meat and the attempted suppression of the scandal. *Lindenstrasse* and other soaps commonly reflect current events such as calls for a boycott of nuclear power.

[4] Cf. Peter Nusser, *Der Kriminalroman* (Stuttgart, 1980), 12.

[5] Biermann's novels include *Potsdamer Ableben*, 4th edn. (Berlin, 1992) and *Violetta*, 3rd edn. (Berlin, 1992), which won the *Deutscher Kriminalroman* award, 1991. Michael Porsche notes that gay and lesbian crime fiction has become an important subgenre in English, as has the 'environmental thriller' represented by the work of Carl Hiaasen and others. 'Journey into the past: Tony Hillerman's *A Thief of Time*', *Amerika Studien – American Studies* 39, 2 (1994), 183–95, here 184–5.

[6] Crime fiction in German mostly imitates English-language models. Writing in a special crime fiction issue of *die horen*, D. P. Meyer-Lenz comments that the pattern of the English 'whodunnit', which focused on the brilliant sleuth, has been superseded in recent decades by Chandleresque fictions designed to function as social and psychological comment, with the ambition of 'illuminating the social contexts' of crime. '200 Jahre Schillers *Verbrecher aus verlorener Ehre*', *die horen*, 31, 4 (1986), 4–6, here 5.

[7] Marion Diedel-Käßner, 'Die verlorene Souveränität des Detektivs', *die horen*, 31, 4 (1986), 24–32, here 27. This and all other German-language sources translated by Tom Cheesman.

[8.] See Doris Lerche, *der lover. Von Männern, Mord und Müsli* (Frankfurt am Main, 1991). Lerche's title is promising but the book has nothing to contribute to the environmental theme – it duplicates the approach of Schatten (see below).

[9.] Viola Schatten's books are: *Schweinereien passieren montags* [*Bad Things Happen on Mondays*] (Frankfurt am Main, 1990); *Dienstags war die Nacht zu kurz* [*On Tuesdays the Night Was Too Short*] (Frankfurt am Main, 1991); *Mittwoch war der Spaß vorbei* [*On Wednesday the Fun Was Over*] (Frankfurt am Main, 1992). References to the last two novels will be indicated in the text by the following abbreviations: *DN* (*Dienstags war die Nacht zu kurz*), *MS* (*Mittwoch war der Spaß vorbei*).

[10.] Friedrich Ani, 'Zu Viola Schatten', *Kultur Aktuell*, Bayerischer Rundfunk, 9 July 1991; Matthias Beltz, 'Mord vor dem Orfeo', *Die Zeit*, 12 July 1991, 59; Gerd Schattner, 'Viola Schatten', *Radiotreff*, Südwestfunk, 30 October 1991; Elke Schmitter, 'Fragwürdigstes. Ein Besuch bei Viola Schatten', *die tageszeitung*, 12 October 1991, 16. Ricarda Huch's *Der Fall Deruga*, first published in 1917, was republished in Berlin, 1980.

[11.] Beltz, 'Mord'. In English, too, 'a striking phenomenon in recent crime fiction has been the attempt at a feminist reworking of the "Hard-Boiled" school of detective fiction'. Rosalind Coward and Linda Semple, 'Tracking down the past: women and detective fiction', in Helen Carr (ed.), *From My Guy to Sci-Fi: Genre and Women's Writing in the Postmodern World* (London 1980), 39–57, here 46.

[12.] See Lerche, *der lover*, 10, and cf. Sabine Deitmer's story, 'Vielleicht hat ihm ein Pferd gefehlt', in her collection *Bye-bye Bruno. Wie Frauen morden* (Frankfurt am Main, 1988), 74–82.

[13.] Similarly, Lerche's dialogue in *der lover* includes the chat-up line: 'Your shampoo smells good. Nettle? Or henna?' (48).

[14.] Schmitter, 'Fragwürdigstes'.

[15.] Arjouni's trilogy includes *Happy Birthday, Türke* (Zurich, 1987; first published Hamburg, 1987); *Mehr Bier* (Zurich, 1987); and *Ein Mann, ein Mord* (Zurich, 1991). The first two are cited from the paperback editions. References will be indicated in parentheses by page references and the following abbreviations: *HBT* (*Happy Birthday, Türke*), *MB* (*Mehr Bier*), *EM* (*Ein Mann, ein Mord*).

[16.] Jochen Schmidt, 'Neue Farben, neue Töne', in Schmidt, *Gangster, Opfer, Detektive: eine Typengeschichte des Kriminalromans* (Frankfurt am Main and Berlin, 1989), 646–58, here 656. See also Hans-Jürgen Fink, 'Marlowes kleiner Bruder in Frankfurt', *Rheinischer Merkur*, 10 January 1992, 21; and Johannes Wendland, 'Der Türke kommt', *Deutsches Allgemeines Sonntagsblatt*, 27 September 1991, 24.

[17.] Fink, 'Marlowes kleiner Bruder'.

[18.] Ulrike Leonhardt, *Mord ist ihr Beruf. Eine Geschichte des Kriminalromans* (Munich 1990), 235.

[19.] Schmidt, *Gangster*, 636.

[20.] See ch.10 of this volume.

[21.] Schmidt, *Gangster*, 657.

[22.] Ulrich Schulz-Buschhaus describes this technique as typical of modern detective fiction with literary pretensions: the solution to the mystery is not offered directly to the reader but must be reconstructed – or it may be implied that the mystery is impossible to solve. *Formen und Ideologien des Kriminalromans* (Frankfurt am Main, 1975), 133.

[23.] *Tatort Umwelt. Kriminalerzählungen von -ky, Michael Molsner, Lydia Thews* (Frankfurt am Main, 1989). A similar collection, republished simultaneously by Fischer TB, features four stories by Swiss writers: *Das Gift im Garten. Kriminalerzählungen von Alexander Heimann, Marcus P. Nester, Friedhelm Werremeier, Peter Zeindler* (Frankfurt am Main, 1989) (first published as *Tödliche Umwelt*, ed. Dietlind Kaiser (Zurich, 1986)). These are very traditional in their approach, in fact more concerned with espionage than with environmental issues, though Werremeier's 'Strix Uralensis' involves the violent demonstrations (which are still continuing) against the transport of nuclear waste. Werremeier has also scripted *Tatort* screenplays on environmental crime: see *Das Gift im Garten*, 239, and Schmidt, *Gangster*, 573.

[24.] -ky, 'Der Krimi aus der Autorenperspektive', in Karl Emert and Wolfgang Gast (eds.), *Der Neue Deutsche Kriminalroman. Beiträge zu Darstellung, Interpretation und Kritik eines populären Genres* (Rehberg and Loccum, 1985), 74–80, here 75.

[25.] Schmidt, *Gangster*, 581–7. See also Michael Bengel, 'Einer soll's gewesen sein. Bemerkungen zu -ky und dem Bindestrichkrimi in Deutschland', in *Der neue deutsche Kriminalroman*, 110–18; and Helmut Schmiedt, 'Gesellschaftskritische Mordfälle', *die horen*, 31, 4 (1986), 51–62.

[26.] References are indicated in parentheses in the text by the abbreviation MN.

[27.] -ky, 'Der Krimi', 75.

[28.] Ulrich Schulz-Buschhaus, 'Die Ohnmacht des Detektivs. Literarhistorische Betrachtungen zum neuen deutschen Kriminalroman', in *Der neue deutsche Kriminalroman*, 10–16, here 15.

[29.] 'Crime fiction texts constantly assure the reader of their mimetic function through continued reference to actual places, dates, people, and organizations. Part of the pleasure of reading depends on this sense of authenticity, allowing me to experience normally inaccessible or forbidden activities.' Sally Munt, 'The inverstigators [*sic*]: lesbian crime fiction', in Susannah Radstone (ed.), *Sweet Dreams: Sexuality, Gender and Popular*

Fiction (London, 1988), 91–119, here 109.

[30.] 'All this might be more bearable if less were made of the attempt to combine hard-hitting realism with literary quality, and if -ky could write a bit better.' Schmidt, *Gangster*, 584.

[31.] Schmidt, *Gangster*, 584.

[32.] -ky, 'Der Krimi', 74

[33.] '. . . das Vehikel, um unsere Gesellschaft zu reparieren'. -ky, cited in Schmidt, *Gangster*, 582.

[34] See Schmidt, *Gangster*, 587–93; Michael Molsner, 'Die Trivialität der Träume. Erfahrungen mit der populären Kultur', in *Der Neue Deutsche Kriminalroman*, 88–98; on Molsner's style, see Diedel-Käßner, 'Die verlorene Souveränität', 30.

[35.] References are indicated in parentheses by the abbreviation ST.

[36.] References are indicated in parentheses by the abbreviation SM.

[37.] Lydia Thews, *Störung der Totenruhe* (Munich, 1984). Reviewed by Werner Jany in *die horen*, 31, 4 (1986), 162–3.

[38.] Rudi Kost, 'Nachwort', *Tatort Umwelt*, 205–7, here 206. See also Schmidt, *Gangster*, 612–17.

[39.] Lydia Thews, 'Frauen als Kriminalautorinnen', *die horen*, 31, 4 (1986), 100–2, here 102.

[40.] A promising newcomer, Sabine Deitmer (cf. note 12), has a female detective who ascribes her headaches to the hole in the ozone layer. Sabine Deitmer, *Kalte Küsse* (Frankfurt am Main, 1993), 78.

[41.] Cf. Peter Nusser's comments on the 'foreigner question' in German crime fiction: 'Das Gastarbeiterproblem im deutschen Kriminalroman der Gegenwart', *die horen*, 31, 4 (1986), 63–9.

[42.] For example Astrid Louven, *Gefährliche Wanderung* (Berlin, 1992), which concerns Kurdish refugees under attack from neo-nazis (and touches on environmental issues too: 15, 37). Peter Freese has published a study of *The Ethnic Detective: Chester Himes, Harry Kemelman, Tony Hillerman* (Essen, 1992), mentioning Arjouni's work (10–11).

[43.] Quoted in 'Lust am Gemetzel', *Der Spiegel*, 1 July 1991, 178.

17 • Apocalypse Nein Danke: The Fall of Werner Herzog

Tom Cheesman

I

Several strands of German ecological thought converge, amid contradictions, in the work of Werner Herzog and in its reception. In Herzog's films, the antagonism between civilization and nature has always been a prominent theme, presented in terms of an apocalyptic imagination. His 'quest for the sublime'[1] leads Herzog to the ends of the earth in search of images of nature which dwarf human beings, while his narratives of revolt – against both society and nature – set linear trajectories of individual desire within circular structures of general futility and doom.[2] Nature is not indifferent, but hostile, murderous; civilization is destructive, driven by heroic but insane Faustian urges to wreck nature; and the human destroyer is finally destroyed by the greater power of the inhuman, non-human 'other'.

Herzog's world-view is commonly dubbed 'romantic', and clearly he draws much inspiration from the writers, painters and musicians of German Romanticism, as well as from the German film-makers of the 1920s and 1930s who also worked in this vein.[3] But Herzog's apparent yearning for a prelapsarian innocence[4] and his sense of universal catastrophe (impending or already upon us) particularly reflect what can be called the eco-apocalyptical tradition, as it developed and became widespread in intellectual and popular culture in and after the 1960s, in many parts of the developed world and most especially in Germany.[5]

Herzog as a 'brand name' – like that of the new German cinema in general – is an international phenomenon; the interlinked development of both, as bodies of work with certain reputations, was driven mainly by reception at film festivals, on campuses and in art-house cinemas outside Germany, particularly in the USA. Eco-apocalyptical thought is also, perhaps by definition, a thoroughly international phenomenon: a matter, often, of thinking globally and brooding locally. Herzog's films represent German and personal perspectives within this wider tendency.

Ironically, though, precisely because he works to find and create images and mythical stories to convey this very popular view of the world, Herzog's films have often been badly received. In part, no doubt, because they are far from popular in form, they have attracted much hostile 'eco-political' criticism, accusing him not only of 'aestheticizing' (and therefore, it is argued, somehow validating) environmental and human catastrophe, but also of being complicit in particular local catastrophes. Though German audiences and critics became notorious for refusing to support the 'new' German film-makers, Herzog suffered more than the usual lack of appreciation, more than the seemingly wilful misunderstanding typically meted out to Fassbinder, Wenders and others – and not only in Germany. The reception of Herzog's work is marked by a wave of enthusiasm among cinephiles (beginning in France and the USA) in the 1970s, followed by a swift disenchantment and, since the débâcle of *Fitzcarraldo* (1982), almost total marginalization everywhere. Eco-political criticism has motivated this marginalization to a great extent. Herzog's reward for exploring the theme of eco-apocalypse with such assiduity has been that he himself has been branded an eco-criminal.

Of course this is not just a matter of shooting the messenger, even if this messenger often seemed to be begging to be shot by apparently committing grave offences against ecological ethics in the course of making his films, and certainly by refusing to apologize. One can ask whether Herzog's brand of apocalypse is not, in the first place, just that: a brand, a marketing device which must be viewed with scepticism, even if it is not entirely cynically promoted.[6] More importantly perhaps, the questions of political naivety, at best, or reactionary (potentially fascistic) irrationalism at worst, need to be addressed. This chapter considers these questions in relation to some of Herzog's feature films and documentaries, including a relatively recent film which perhaps reawakened the memory of Herzog in some quarters: *Lessons in Darkness: Requiem for an Uninhabitable Planet* (1992), a poetic documentary filmed in Kuwait in October 1991.

II

For a decade, *Aguirre, the Wrath of God* (1972), *The Mystery of Kaspar Hauser (Every Man for Himself and God against All)* (1974), *Stroszek* (1976), *Nosferatu* (1978, commercially his most successful work by far) and *Woyzeck* (1978) made Herzog a darling of international cinephiles. But no sooner had he gained this following than he began to lose it. *Fitzcarraldo* marked the turning-point: with his biggest-ever budget, Herzog created a political furore even in the pre-production phase by sparking a small civil war in the Peruvian Amazon. Since the completed film was not his greatest masterpiece, nor commercially successful, and since it could not provide a clear vindication of the violence engendered in the process of its creation, his reputation has never recovered. With the eco-fiction *Where the Green Ants Dream* (1984), and then the slavery epic *Cobra Verde* (1988), Herzog was seen as merely treading water, repeating himself less eloquently.[7] The mountain movie *Scream of Stone* (1991), his last feature film, was universally panned, if not simply disregarded.[8] Much has been written about him, but unlike his compatriots in the new German cinema (Fassbinder, Kluge, Schlöndorff, Wenders et al.) he is now ignored in Germany and seldom more than marginally acknowledged in other writing on recent European cinema. Only one book in English is devoted to serious study of Herzog's films, and the editor and all but a couple of the contributors 'betray . . . a manifest hostility' to the person, his projects, his aesthetics and ideology – or those which are ascribed to him.[9]

Of all the film-makers who came to critical attention in the glorious years of the new German cinema, Herzog stands out as one who played the game of mythic self-promotion, at first with outstanding success, only to lose at last. Presenting himself less as a visionary (Romantic) artist, a poet-priest (though this is how many of his fans perceived him), than as a self-sacrificing artisan in the service of an enigmatic art, and as a heroic athlete, whose film projects involved immense personal risks to life and limb, Herzog ensured that his work made headlines in many cases even before the films were in the can. In particular, he repeatedly provoked debate about the damagingly exploitative manipulation not just of performers, but also of exotic places and their inhabitants, landscapes and cultures.

Nina Gladitz arguably ensured that this debate acquired a certain public insistency by pursuing a 'shrill vendetta'[10] against him, in a collection of documents entitled *Der Fall Herzog* (n.d.)[11] and in her film, *Land of Bitterness and Pride* (1984). Gladitz set Herzog beside Leni Riefenstahl, in connection with the latter's alleged Nazi sympathies and exploitation, in pre- and post-war work, of gypsies and Marsh Arabs. Herzog stands accused of mistreating the midgets in making *Even Dwarves Started Small* (1970), the profoundly deaf, dumb and blind in making *Land of Silence and Darkness* (1971), the 'part-time schizophrenic'[12] Bruno S. in making *Kaspar Hauser* and *Stroszek*, an entire cast subjected to hypnosis in making *Heart of Glass* (1976), all the citizens of Delft and some ten thousand white rats (dyed grey) in making *Nosferatu* – and of course, most notoriously, the Jivaro and Aguaruna Indians and their environment in the Peruvian Amazon jungle, in making *Fitzcarraldo*.[13]

In 1984 Herzog taunted his politically correct critics by documenting the oppression of the Miskito Indians of Nicaragua by the Sandinista regime, in *The Ballad of the Little Soldier*: this provoked renewed allegations of manipulation and exploitation of his subjects, of political naivety or reactionary malevolence. At a time when many in Herzog's natural audience (young, more or less left-wing, art-house cinema-goers) were stocking up on Nicaraguan coffee and wearing campesino shirts, this was not unlike Bob Dylan's outburst at the 'Live Aid' concert, calling on the global audience to remember the plight of small farmers in the American Midwest: an unforgivable affront to the soft liberal consensus. *Where the Green Ants Dream*, released in the same year, was more grist to the mill of eco-political critique. It was regarded not only as falling between the stools of ethnographic-environmentalist documentary and exotic fiction, but also as showing, above all, a manipulative, egomaniac director who 'moves the Aborigines before his camera in much the same way as the mining company threatens to have them moved by the police'.[14]

Herzog depicts the Aboriginal tribespeople as heroically resisting the catastrophe of 'progress' in the shape of the destruction of the landscape they inhabit, but he allows them no inner life except for – as Thomas Elsaesser puts it – 'a fantasy concoction of his own making' (the cosmic mythology of the 'green ants' is Herzog's invention) or else an objectivizing assertion of 'the physiological-anthropological affinity the heroes are supposed to entertain with their environment'.[15] Likewise, his depiction of the inhabitants of the North African desert in the epic, poetic documentary *Fata*

Morgana (1970) subordinates their experience of life to the landscape in which they live, making them mere tokens of a metaphysical suffering shorn (to his detractors' annoyance) of any specific historical, social, economic or political context. They become components of a desolate landscape which itself represents the failures both of creation, and of civilization.

So it is alleged. But Elsaesser also observes that the inhabitants of deserts and jungles in Herzog's films 'look at us, the white spectators, and seem to be looking through us'; Herzog's use of cinematographically 'primitive' techniques (especially the stationary, frontal full shot of a person gazing into the camera) allows him to 'let his characters let us know that they perform, and that their performance has a gravity beyond tragedy'.[16] Herzog's use of such people is commonly seen as ethically problematic to the extent that he neglects to declare his own interests subordinate to theirs. But he never betrays their dignity by committing the alternative and much more widespread error, which is to pretend to speak on their behalf.

Herzog's sombre, melancholy *Kulturpessimismus*, his repeated invocations of existential despair, his representations of doomed Faustian (or Hitlerian) heroes in struggle against the hostility of external and human nature (his Kinski characters), or of isolated misfits facing an uncomprehending, 'mad' social order (his Bruno S. characters) – these features have become (again) deeply unfashionable. The mesmeric, haunting beauty of his images (accompanied by seductive music or incantatory voice-over), his uses of mythic narrative, and his erasure of the distinction between fiction and documentary were suspect in the politicized 1960s – Herzog spent his first ten years of film-making without gaining any recognition – and they became suspect again in the 'postmodern' 1980s. Herzog's *œuvre* circles around a restricted set of themes – centred on the individual's futile revolt against the hostility both of the natural and the social world – and deploys a restricted set of techniques to convey them. Thus he runs the risk of boring an audience on the look-out for novelty. But it is the perfection of each of his works which is most troubling to postmodern art-house audiences, who seek space for the play of interpretation in unresolved tensions between filmic and pre-filmic elements. Herzog leaves no such space: his films clearly mark out the limits of their own interpretability, pointing beyond themselves quite unplayfully to the ultimate mysteries of life, death and the universe.

The antagonistic relation between humankind and nature – the environment, the animal, vegetable and mineral world – is among Herzog's preoccupations from a very early stage in his work. One of his first short documentaries, *Spiel im Sand* [*Game in the Sand*] (1964), has never been shown in public: it depicts a group of children playing with a cockerel, which they bury up to its neck in sand, and proved to be a typical example of Herzog's 'documentaries' which 'do more than record events: they invent them.'[17] The film apparently 'became a terrifying document of violence', as the children's latent sadism was unleashed during the filming.[18] Here, *in nuce*, we face the ethical issues of manipulation and destructive intervention which Herzog's critics have repeatedly raised, in this case regarding both the creature which was tortured and killed, and the children who may perhaps have been traumatized by the experience of losing their civilized inhibitions. That Herzog has not released this film indicates at least that he is well aware of an ethical boundary and a taboo.

He thematized the issues directly in *Precautions against Fanatics* (1969), a film made on a German race-course. Trainers and owners are shown drilling, doping and otherwise treating the horses as objects, while declaring that they have their best interests at heart and repeating that the animals needed to be 'protected against fanatics'. Meanwhile an unidentified, one-armed old man constantly invades the frame proclaiming that he alone is the creatures' true spokesman. Eric Rentschler comments: 'In essence, the film champions the natural against its tamers, preferring the fanaticism of the eccentric old man who speaks a broad Bavarian to the mercenary instrumentalism of the trainers.' Rentschler places this laconically comical film at the start of a regrettable development in Herzog's work and ideology, in which he effectively changes sides: from identifying with the social outsider who really is 'in touch' with nature-in-itself, and condemning the official guardians and formal owners of nature-as-object, in the course of his career Herzog 'has come to intrude upon and often abuse . . . vulnerable entities'.[19]

When *Lessons in Darkness*, depicting the landscape of Kuwait in the aftermath of the Gulf War 'as if it were seen by a traveller from another world witnessing the destruction of ours',[20] was premièred at the Berlin Film Festival in 1992, it provoked predictable responses:

[The film] consisted largely of highly aestheticized shots of burning oilfields ... set to a score of classical music – primarily Wagner. In an appearance before an incensed (but small) gathering after the showing, the director was in his best authoritarian form, retorting to complainers, 'I don't care if you don't like the music, I like it that way.' If any doubts about his aesthetic politics remained, performances like this clearly exhibit a purposeful self-positioning as an artist in search of pure images above mere mortal concerns.[21]

Amos Vogel, one of Herzog's few remaining supporters, describes the same press conference differently. The 'great and beleaguered German director' faced criticism on the grounds of 'not identifying Kuwait, not condemning Saddam Hussein (or Bush), not stating outright that this film was anti-war (every single frame is)', as well as for succumbing to 'aestheticism':

Herzog in his answers did not stoop to their level ... He even (quite properly) admitted to the 'beauty' of the fires ... and of atomic explosions. He stated that he deliberately did not want to name Kuwait, since the point of this film was to go far beyond this war and its contestants. His intention, he concluded, in making this film had been to preserve these horrifying, never-before-seen images for the memory of mankind.[22]

Lessons in Darkness thus typifies the difficulty created by Herzog's work for politically minded critics: he addresses basic issues in contemporary global politics – environmental destruction, war, (neo-)colonialism and oppression – but views them *sub specie aeternitatis*. If his ideal audience is 'the memory of mankind', that can lead to individual human beings, at present, feeling a certain resentment at the director's arrogation of god-like powers. It perhaps also suggests an ambiguity: does Herzog imagine the human species remembering these images (and his work in creating them), or does he imagine mankind as an extinct species, remembered by some other consciousness?

III

Herzog's quest for the sublime has repeatedly focused on landscapes, and in the years of his fame he was frequently quoted on the subject of landscapes 'embarrassed' (his own translation of *beleidigt*) by human presence, implying the destruction of pristine nature wrought by civilization.[23] Francis Ford Coppola is reported to have told people to watch Herzog's *Kaspar Hauser* in order to learn how to film landscape.[24] (And *Apocalypse Now* (1979), as a film about a river journey through jungle, was directly modelled on *Aguirre*.) Herzog's fascination with landscape, with the aesthetics of nature, distinguishes him from all his peers in the new German cinema. It has been pointed out that Herzog alone uses 'absolute landscape': images of natural scenery which are neither motivated by the point of view of a character in a film, nor possess any denotative meaning. Instead they function as the film-maker's direct address to the viewer, displaying nature as 'an elemental force' inviting awed contemplation.[25]

In these scenes, any human activity, or traces of human activity, appear as violations or as tragic absurdities. The desert docu-epic *Fata Morgana* is composed almost entirely of such sequences, as is *Lessons in Darkness*. Other instances are the sequence of images reminiscent of Caspar David Friedrich, accompanied by a chorale by Orlando di Lassus, which we see while Kaspar Hauser sleeps on the hillside after being brought out of his cell; or the images of clouds and mountains and sea, suggesting world apocalypse, accompanied by the music of Popol Vuh and the doom-laden prophecies of the shepherd, Hias, which open, punctuate and close *Heart of Glass* (1976) – a film which tells an allegorical tale about the self-destruction of industrial civilization, based on Wilhelm Hauff's *Das kalte Herz* (literally, *The Cold Heart*).[26]

But Herzog's Romanticism regarding landscape is not simply a matter of sentimental longing for Eden. His own fall from grace at the time of *Fitzcarraldo* was hastened by his very unromantic and oft-repeated remarks about rainforests, which struck a highly iconoclastic tone, coming just as the popular movement to save them from the depredations of civilization was getting under way: 'The jungle is . . . obscene. I always make this speech against Mother Nature. It is full of pain, full of . . . full of . . . collective murder. That is the common denominator, murder. It is what makes the jungle the jungle [second and third ellipses in the original].'[27] Appealing to his personal experiences of hostile environments, Herzog vehemently rejects any sentimentality about nature. The 'unembarrassed' landscape is evidently one which is so wild that if humans have ever encroached upon it, their traces have been erased by hostile elemental forces.

For the documentary short *La Soufrière* (1976),[28] Herzog took two cameramen to the island of Guadaloupe, which had just been evacuated because a major volcanic eruption was imminent. In fact (of course), to the amazement of vulcanologists, the volcano did not erupt; otherwise there might have been no film to watch. What we see is a record of futile, disappointed waiting for the 'inevitable catastrophe' in a half-wild landscape – a road leads up the mountain, but the crew can only go so far before clouds of volcanic fumes threaten to suffocate them, and they hastily retreat. This falls bathetically short of the Byronic gesture – Manfred shaking his fist at the elements on the brink of the precipice – which, as Peucker remarks, is a common 'Romantic attitude' in Herzog's films.[29] Suspense is, however, maintained by helicopter shots of the threateningly smoking crater, accompanied by bombastic Romantic music; an exploration of the creepily deserted town of Bas-Terre; a historical mini-documentary about the eruption of 1802 which killed several thousand people in Martinique (using 'before and after' photos of the town); and Herzog's hushed, tense commentary, full of feverish anticipation of the crack of doom ('we had the impression that these were the last hours of this town and the last pictures ever taken of it'). To fill the time while nothing happens, they seek out a few old men who have refused to leave the island. They prove to be largely unresponsive subjects, with nothing heroic about them and nothing interesting to say about their calm indifference to fate and death. Herzog's overdubbed translation of their patois tries and fails to ascribe to them a stoic philosophy. It seems likely that their main reason for remaining on the island was sheer passivity. Towards the end, as it becomes clear that the inevitable catastrophe will not take place, Herzog comments – over the now ironic counterpoint of swelling dramatic chords – that he and his project have become 'pathetic' and 'embarrassing'.

This comical failure to fulfil a titanic ambition falsifies Herzog's frequent claim to be able to 'direct landscape'.[30] The mountain refuses to perform. But Herzog's clear sense of his own absurdity could lead one to wonder whether the general tendency to convict him of reactionary, Romantic idealism is not wide of the mark. *La Soufrière* exists as a film in place of two others which might have been made, had the volcano erupted: one, a terrifying and thrilling film about the eruption, probably involving some injuries, even deaths, among the men on the island, but including a narrative of escape from death by at least one of the crew; the other, a film buried along with all of them, which no one would ever see. With this in mind, it becomes apparent that Herzog's theme is human survival on this inimical planet, and the costs it involves. His awesome landscapes keep asking: can we live here?

IV

Another of Herzog's oft-repeated phrases invokes the fear of humanity's extinction and suggests that film-makers have a crucial role to play in averting it:

> What we have around us are worn-out images, and this is a very danger-ous thing. We have become aware of certain dangers that surround us. We have understood for example that nuclear power is a certain danger, we have understood that overcrowding of this planet is a very grave danger, maybe even the biggest. We have understood that all the destruc-tion of the environment is an enormous danger that we face, and I truly believe that the lack of adequate images is a danger of the same magni-tude. If we do not develop adequate images we will die out like dinosaurs.[31]

At first sight, this is certainly not a rational view: whatever stands between the present condition of the planet and the possibility of its becoming a monument in space to our capacity for destruction, it is surely not going to be the work of independent film-makers. Here the eco-apocalyptic imagination seems merely to be put in the service of an apology for Herzog's profession and obsessions.

The question of the value of rationality is an enigma that Herzog's work repeatedly confronts. The accusation that he 'instru-mentalizes' others, and the general other of humanity, that is nature, for aesthetic ends – and does so using the part-artisanal, part-industrial apparatus of cinema – turns against him his own critiques of modern, technological civilization based on rationality. Herzog tends towards magical thinking – most famously he claims to have prevented Lotte Eisner from dying in Paris, in winter 1974, by undertaking a pilgrimage, on foot from Munich, to her bedside.[32] But it is not clear just how seriously he takes himself when he tells these kinds of stories. In *Kaspar Hauser* various spokesmen of bourgeois rationality and logic are comically and very effectively defeated by the hero's obdurate recourse to 'unreasonable' intuitive thinking. But when Kaspar Hauser narrates his dreams, he tells of a series of discrete images (of exotic and sublime landscapes) that preface a story he cannot remember, or never knew: perhaps this indicates the severely limited usefulness of the pure imagination, prior to reason. The rejection of rationality is taken as the most serious affront by Herzog's critics, but oddly, even his greatest detractors keep finding that he 'unintentionally' *does* put a value, even a decisive value, on rationality.

Heart of Glass stands as one of Herzog's most provocative experiments in obfuscation. As a pastiche Bavarian *Heimatfilm* ['homeland film', normally a celebration of German rural virtues], it allegedly betrays his sympathies with the reactionary ideology of real *Heimatfilme*. A three-page summary of the film by Kraft Wetzel displays typical critical hostility; in the following citation Wetzel describes the closing sequence, which depicts a vision by the shepherd-seer, Hias. The sequence was shot (mainly from helicopters) at the Skellig Islands, off the west coast of Ireland:

> Two craggy islands rise out of the sea, 'on the last edge of the inhabited world'; the knowledge that the earth is round has not reached them yet. After spending years staring out over the sea, two men dare to risk everything: they set off in a rowing boat to find the edge of the world, braving the danger of being swallowed up by it. Herzog has Hias denounce their departure as 'pathetic and senseless', and adds a closing title that almost mocks them: 'They may have perceived it as a sign of hope that the birds followed them to the open sea.' But the power of these final images, which Herzog intended as one last concentrated expression of his Romantic pessimism about civilization, escapes his own grasp: with the image of the bold men in their tiny boat, quite contrary to his intentions, he has managed to create one of the most impressive and moving metaphors for the pathos of [the] Enlightenment, for the escape from ignorance.[33]

Similarly, much of Rentschler's critique of this film depends on the assumption that Herzog only accidentally or unconsciously created one of its key features, namely the 'doubling' that identifies Hias, the prophet from the mountains who is apparently the film-maker's mouthpiece, with the factory owner who becomes a murderer in his search for the lost secret of ruby glass, and who (Rentschler claims) is 'really' Herzog's representative within the film. Moving on to observe that the figure of Fitzcarraldo fuses 'the visionary and the entrepreneur', Rentschler alleges that *Fitzcarraldo* provides the 'most chilling expression' in Herzog's work of the 'triumph of aesthetics over all other considerations'.[34]

Such attempts to pin guilt on Herzog for countenancing – at the extreme – murder in order to create works of art that are meant to be seen as 'beyond morality and history'[35] (and, of course, beyond politics) read the director's intentions somewhat naïvely in the behaviour of his films' protagonists. We face here exactly the problem faced by critics of Goethe's *Faust*, in dealing with a criminal tragic hero and representative of modern western man in general (colonialist–imperialist–exploiter–destroyer–murderer), with whom however the reader (or viewer) identifies, and who in the end is 'redeemed' within the text.[36] Both *Faust* and *Fitzcarraldo* end with triumphant opera. The real questions, though, are: do these finales erase our ability to remember what we have seen and heard, and does the identification with the hero, which these texts promote, preclude critical distance? If we can answer in the negative, surely we must admit that Herzog has taken great risks – not least with his own reputation – in order to create films which invite the viewer to reflect upon the conditions and consequences of various kinds of human ambition, in the context of the questions of survivability which beset us in the atomic age.

In *Lessons in Darkness*, Herzog made a science-fiction film about the end of the world out of a documentary about post-war Kuwait. As regards 'the politics of vision', he supposedly – according to John E. Davidson writing in *New German Critique* – only in 'a weird twist of fate', and through his individual and guilty 'fascination with catastrophe and his drive to create essences', brought images to the screen which show what the Gulf War actually brought about, and thus quite 'inadvertently . . . restor[ed] vision to the official, literally visionless images of the war'.[37] The strict control, indeed lack of images from the Gulf War was widely commented upon: it was a 'virtual war', to the point that the most notorious postmodern philosopher, Baudrillard, was able to maintain that it never 'really' took place at all.[38] The images made available to television audiences were largely indistinguishable from those displayed in video war-games. Herzog's critics assume that, in making this documentary, the film-maker remained naïvely unaware of, or else sublimely indifferent to the fact that his 'never-before-seen' images inevitably have political content, since they are representations that have escaped a very powerful regime of media censorship. They assume this, merely because his spoken commentary and his method of editing images and soundtrack fail to declare a political viewpoint in terms of assigning particular responsibility for the catastrophe he depicts.

Of *Where the Green Ants Dream*, Herzog wrote that the film shows little of the diversity of Australia ('it would have made the film too long') or of the struggle and political activity of the tribal Aboriginals: 'I did not feel it was necessary to show more, because this ... is well publicized.' He goes on to discuss the tensions between tribal and urban Aboriginals, but 'I did not want to get into that side of the problem because I wanted to make a simple and clear film.'[39] Herzog's ambitions effectively marginalize all sorts of other points of view, and he is thoroughly aware of that fact. He takes for granted the existence of other discourses, filmic and non-filmic, official, minority and oppositional. His films, as interventions in debates, seldom trouble to acknowledge other voices. But given the actually rather marginal status of his own discourse, it seems perverse to describe him as 'authoritarian' merely because he seeks to retain control over his utterances.

Davidson's critique of the politics of the aesthetic in *Aguirre* and *Lessons in Darkness* finally, just like Wetzel's comments on *Hearts of Glass*, arrives at a view of Herzog as a radically provocative filmmaker serving enlightening ends. But the critic's rhetoric cunningly denies Herzog the credit for any such achievement. Again, the filmmaker means to be a reactionary obfuscator, but fails, and unintentionally becomes progressive. In a key phrase: 'Like *Aguirre*, the point when Herzog succeeds is the point at which he must necessarily fail: the West tries to distance itself from his tactics and his depictions, because they unabashedly enact and depict the violence of the West.'[40] Or, concluding Davidson's essay:

> Herzog glories in the 'aesthetic beauty' of the world in which those oilfields burn, and has no qualms about showing that domination, destruction and death of 'others' provide the foundation for aesthetics. Thus, the hermeneutical impasse posed by Herzog's films ... [challenges] our ability as critics to recognize that not merely the misuse of the aesthetic, but its very (re)production furthers the causes of domination, which we claim to subvert.[41]

Thus, Davidson sees further and better than Herzog (not to mention
Rentschler et al.) into the heart of darkness of *all* contemporary art,
without exception.

This would be merely a tedious instance of critics competing, in
their circus of jargon, for the badge of 'most radical' – were it not
that there is a genuine problem in discerning the political ramifica-
tions of the eco-apocalyptical imagination. Especially in the German
context, in view of the continuities within 'green' thought before,
during and after the Hitler period, the critique of modern civilization must be suspected of harbouring the wish for an authoritarian
and inhumane solution to its discontents.

V

The history of the apocalyptic imagination, from the second century
BC to the present, is sketched by Hartmut Böhme in an essay on
'The past and present of apocalypse' (1988). The essay takes as its
starting-point Derrida's well-known satirical comments on the
'apocalyptic tone' in culture and philosophy in the 1970s and
1980s,[42] but Böhme argues that apocalypse is historically 'a root of
critique – indeed, it *is* critique – in those forms which, contrary to
the "rational" speech of the privileged [*der Vornehmen*], the rulers,
the discursively powerful, use poetic images which are accessible to
the oppressed and persecuted'.[43] Apocalypse does not mean the
'final end' of things, but catastrophic revelation of the truth and a
new beginning: the tradition of apocalyptic imagining serves to
control and assuage the fears of those who suffer. It is 'a metaphysical strategy of self-preservation at the frontier of absolute
negation, the abyss of nothingness which threatens all human
history'.[44] So it offers a hope of redemption which is 'élitist' only
in the sense that it promises the oppressed that they shall enjoy
privileges denied to them at present: the last shall be first, justice
shall rule. But since the Book of Revelation has gone increasingly
unread,

> In face of the bestial aggressivity of the vision, it was forgotten that this
> was aggression against aggression; in the face of the bloodlust of
> absolute war, it was forgotten that this was a war against war; in the face
> of the savage natural catastrophes, it was forgotten that they were

directed against the man-made catastrophes; in the face of the orgiastic destructions, it was forgotten that they served to bring about paradise: no hunger and thirst, 'all tears' wiped away 'from their eyes', longing fulfilled by the 'living fountains of waters' (Revelation 7:16–17).[45]

Böhme contrasts the apocalyptic discourse, rooted in Judaic tradition, with the discourse of 'the Platonic tradition all the way to Habermas', which supposes that truth can be arrived at through rational conversation, and thus is predicated on the exclusive participation of 'subjects unburdened by the need to act, well-fed and peaceable members of the leisure class, inhabitants of the earthly paradise' (a model of truth-finding which 'even today denies its silent dependence upon slaves and the voiceless exploited'). Truth in apocalyptic discourse, on the other hand, is

crude, barbaric, born of need, explosively sprung out of dumb existence, uncontrolled, radical and ungoverned by rules, never conceptual but plummeting away in cascading streams of poetic images, barely finding its grip in always implausible magical-aesthetic forms, perfectly uninhibited towards the disciplines of dominant discourse, never reflective and cool, but thrown hither and thither between the extremes of feeling of which people are capable: revenge, hatred, rage unleashed, bestial cruelty and the most tender gestures of love, passionate self-sacrifice and pleasure in death.[46]

This is indeed the very 'other' of rationalism; and it is highly evocative of the content of Herzog's work – if we can overlook the director's highly sophisticated, controlled and reflective use of aesthetic forms.

Böhme goes on to tell an unsettling story about Kant's 'invention' of the sublime as a key modern aesthetic category: in the course of this story, the clear-cut distinction between apocalypse and reason becomes untenable. In the context of cultural secularization, as the notion of redemption in a 'second creation' became merely fictional, the apocalyptic imagination was freed from its moorings in the discourses of organized religion. The often plebeian fears it had controlled now threatened to run amok, subverting the project of the Enlightenment, of modern, bourgeois, capitalist development. Kant's model of the sublime (Böhme claims) was therefore devised to reassert the power of the rational subject over the terrors of destruction by elemental natural forces, or the ravages of war – by powers utterly exceeding those of real individuals, their imaginations and their powers of depiction. It did so by staking an absolute claim on behalf of reason itself. The affirmation of pure reason survives the destruction of the body, possibly of every body: 'even in the total catastrophe that exceeds every human imagination and annihilates real humanity, all that counts is whether the catastrophe can be *thought* in such a way that in it, through it and beyond it, reason triumphs.'[47] Thus disembodied, in this 'cognitive psychodrama' reason 'recycles fear in the aesthetic combustion process of the sublime'.[48] Terror can become pleasure. And so, Böhme argues, in modern history the secularization of apocalypse as the aesthetic sublime becomes the idealistic impulse which celebrates war, the destruction of the products of nature and technology alike, as sublimely great: 'The aesthetic of the sublime, as a structure which is at work in today's scenarios of human species catastrophe [*den anthropogenen Katastrophenszenarien*], is the temptation to escape from the fear of the end by bringing it about oneself.'[49] The end-product of the pursuit of the sublime within the rationalist tradition must be 'to send the earth through silent space as a grandiose monument to our own superpotency': 'One can think of no more sublime work of art than the abandoned, silent earth.'[50]

VI

Herzog's position in relation to this catastrophic history of catastrophic thought, and the actions it has engendered, seems to be appropriately ambivalent: he can vindicate neither what passes for rationality, and the damage it has done; nor the irrational, even magical hope of survival, and of the damage being repaired; but he is also unwilling to abandon the effort to find ways beyond this impasse.

In *Lessons in Darkness*, amid the apocalyptic landscapes, he shows only one individual: a mother who has lost her power of speech after witnessing the torture and killing of two of her sons. With her surviving youngest son in her arms, she stares into the camera, seemingly unseeing. 'In a pathetic attempt to tell what happened, [she] utters a few horrifying, animalistic sounds while her eyes express the horror she cannot. It is an unbearable scene. It does not end.'[51]

Twenty years earlier, Herzog had made *Land of Silence and Darkness*, an extraordinary feature-length study of a group of profoundly deaf, dumb and blind people in a German hospice, which succeeds in making the viewer appreciate the mystery of their experience and gives them enormous dignity as individuals. In the last scene, one of them walks slowly into a garden and up to a tree, which he gently embraces. The scene is shot from some distance; we do not intrude on the man's communion with nature; but the shot is held for several minutes. As Herzog has remarked, without the preceding ninety minutes of documentary the scene would be simply meaningless or trite; he can justifiably claim that, in the context both of information and of aesthetic structure that he created, it becomes 'one of the deepest moments you can ever encounter in the cinema'.[52] The touch between the tree (German culture's icon of all external nature)[53] and the speechless, sightless, unhearing man wonderfully expresses the hope, born out of tragedy, for universal reconciliation between people and nature.

The speechless Kuwaiti mother represents the entirety of suffering humanity in the face of the real presence of apocalypse. Around the short sequence in which she appears, Herzog orchestrates visions of the end of the world. The imaginary journey is punctuated by intertitles such as 'The war', 'After the battle', 'Finds from torture chambers', 'Satan's national park'. At the film's long-drawn-out climax, we are taken into the sublime infernos of burning oil wells shrouded in immense clouds of fumes which block out the sun, where every shot is filled with 'orgasmic fires', 'eruptions of a primal, world-shattering force,' of 'a demonic virulence that calls into question our existence on earth': 'we are forced to view this event as a catastrophe of unimaginable dimensions and consequences, one that raises a question mark over the entire human race.'[54] The camerawork, slow-moving throughout, often shot from helicopters or else from the heat-shielded tractors used by the crews working to cap the burning oilrigs,[55] is accompanied sometimes by a sparse, cryptic, ice-cold commentary, occasionally quoting the Book of Revelation ('I am so weary of sighing, O Lord, make that the night cometh'), but mostly by the music not just of Wagner, but of Mahler, Verdi, Grieg, Prokofiev, Pärt and others besides.

In *Fitzcarraldo*, 'opera seemed to provide alternatives to the destruction of nature'; but here, the apocalyptic landscapes 'remain mute, indeed even silenced by musical thunders from *The Twilight of the Gods*', opines Koepnick.[56] But it is not music that silences these landscapes: war did that. Rather, music – coupled with the extremely artful composition and editing of the shots – makes it possible to carry on looking at them, by creating visual and aural beauty out of chaos and terror. Herzog's vision of apocalypse goes beyond the aesthetic sublime, because it attempts to imagine and depict, rather than reason about the final cataclysm. As his quotations from the Book of Revelation suggest, he is in part seeking to revive the pre-modern tradition of apocalypse. This seems reactionary only until we remember that apocalypse is a popular, critical tradition, maintaining hope in the face of extremes of suffering. Herzog lacks pre-modern faith; hope remains mute. It can only be said to exist – but then it *can* be said to exist – if the film still has a human viewer to be moved by the mother's despair. And this depends on the continued survival of a modern, technological civilization which does not entirely forget Herzog, and all the others who point out what that civilization costs in human and ecological terms. The vindication of Herzog – and also the reason for his fall from grace with the fickle art-house cinema public – lies in the fact that he crafts astonishing films about the last things, without pretending to know what comes after.

Notes

[1.] Brigitte Peucker, 'Werner Herzog: in quest of the sublime', in Klaus Phillips (ed.), *New German Filmmakers* (New York, 1984), 168–94.

[2.] A clear introductory analysis of Herzog's themes and techniques, focusing on his first feature, *Signs of Life* (1967), and *Aguirre*, is provided by Dana Benelli, 'The cosmos and its discontents', in Timothy Corrigan (ed.), *Between Mirage and History: The Films of Werner Herzog* (New York and London, 1986), 89–104.

[3.] Brigitte Peucker, 'Literature and writing in the films of Werner Herzog', in Corrigan (ed.), *Between Mirage and History*, 105–18; Judith Mayne, 'Herzog, Murnau and the vampire', ibid., 119–32.

[4.] 'But . . . *Fata Morgana* . . . gives an ironic lie to the narrator's nostalgic wish for a return to a moment of cultural innocence' and 'the myths of transcendence proffered in [Herzog's] films are rendered radically incongruous with the spectatorial position the films produce in the viewing': Alan Singer, 'Comprehending appearances: Werner Herzog's ironic sublime', in Corrigan (ed.), *Between Mirage and History*, 183–205, here 188. Singer's argument seems to be similar to that which I am advancing, but he 'at times presents an impenetrable wall of abstractions' (Christopher J. Wickham, reviewing Corrigan's book in *Film Criticism* 12 (1987), 31-6, here 33).

5. For the German context, see ch.9 in this volume. For international perspectives, see for example David Dowlby, *Fictions of Nuclear Disaster* (Houndsmill and London, 1987); Lois P. Zamora, *Writing the Apocalypse: Historical Vision in Contemporary US and Latin American Fiction* (Cambridge, 1989); Steven Goldsmith, *Unbuilding Jerusalem: Apocalypse and Romantic Imagination* (Ithaca and London, 1993).

6. Thomas Elsaesser, 'An anthropologist's eye: *Where the Green Ants Dream*', in Corrigan (ed.), *Between Mirage and History*, 133-58, describes Herzog as 'a one-man cinematic genre' (133).

7. But see the vigorous defense of *Cobra Verde* by Herbert Golder, 'Herzog and the demons', *The Guardian*, 29 July 1988. Golder compares the film to Euripides' *Trojan Women* as a study of the human consequences of slavery.

8. For example: Nigel Andrews, 'Lure of the wanderlust movie', *The Financial Times*, 3 November 1992.

9. Christopher J. Wickham, reviewing Corrigan (ed.), *Between Mirage and History*, 33. It is symptomatic that in the series of paperback studies of directors published by Carl Hanser Verlag, no. 9 (1976) is devoted to three radical directors as a group, rather oddly including Herzog, and no. 22 (1979) to Herzog singly – since when German film scholarship has comprehensively ignored him: *Herzog/Kluge/Straub. Reihe Film 9. Mit Beiträgen von Ulrich Gregor [et al.]* (Munich and Vienna, 1976); *Werner Herzog. Reihe Film 22. Mit Beiträgen von Hans Günther Pflaum [et al.]* (Munich and Vienna, 1979).

10. Rentschler, 'The politics of vision: *Heart of Glass*', in Corrigan (ed.), *Between Mirage and History*, 159-82, here 169. Though Rentschler distances himself from Gladitz here, he essentially supplies scholarly powder to her cannon: his critique builds on a comparison between *Heart of Glass* and Riefenstahl's *The Blue Light* (1932).

11. Nina Gladitz, *Der Fall Herzog. Eine Dokumentation der Gesellschaft für bedrohte Völker. Reihe Vierte Welt Aktuell Nr. 12, Sonderausgabe* (Freiburg, n.d.). I am grateful to Theodor Rathgeber of the Gesellschaft für bedrohte Völker [Society for Threatened Peoples], Göttingen, for supplying this reference.

12. Robert P. Kolker, *The Altering Eye: Contemporary International Cinema* (Oxford, 1983), 263.

13. On the violence and the controversy surrounding the making of *Fitzcarraldo*, see Gladitz's film and *Der Fall Herzog*; Les Blank's film, *Burden of Dreams* (1982); and Les Blank and James Bogan (eds.), *Burden of Dreams: Screenplay, Journals, Photography* (Berkeley, 1984). This

particular case against Herzog is invoked by many of the contributors to Corrigan (ed.), *Between Mirage and History*. The editor's introduction is relatively even-handed ('Producing Herzog: from a body of images', 3–22), but he still describes Herzog's work as 'annoying', annoyingly often. The issues surrounding Herzog's representation of, and alleged complicity with, colonialist oppression and ecological destruction are addressed by Victoria M. Stiles, 'Fact and fiction: nature's endgame in Werner Herzog's *Aguirre, the Wrath of God*', *Literature/Film Quarterly*, 17 (1989), 161–67; at greater length by John E. Davidson, 'As others put plays upon the stage: *Aguirre*, neocolonialism, and the new German cinema', *New German Critique*, 60 (1993), 101–30, and by Lutz Koepnick, 'Colonial forestry: sylvan politics in Werner Herzog's *Aguirre* and *Fitzcarraldo*', *New German Critique*, 60 (1993), 133–59. Herzog puts his side of the Fitzcarraldo story in Andreas Rost (ed.), *Werner Herzog in Bamberg. Protokoll einer Diskussion, 14.–15. Dezember 1985* (Bamberg, 1986).

[14.] Elsaesser in Corrigan (ed.), *Between Mirage and History*, 149.

[15.] Ibid., 146.

[16.] Ibid., 156.

[17.] William Van Wert, 'Last words: observations on a new language', in Corrigan (ed.), *Between Mirage and History*, 51–72, here 71.

[18.] See Emmanuel Carrère, *Werner Herzog* (Paris 1982), 11.

[19.] Rentschler in Corrigan (ed.), *Between Mirage and History*, 176.

[20.] Amos Vogel, 'Lessons in darkness', *Film Comment*, 28 (1992), 69–70, here 70.

[21.] Davidson, 'As others put plays upon the stage', 129.

[22.] Amos Vogel, 'Lessons in darkness', 70.

[23.] '"I want to find some landscape," he said, "that has not become embarrassed". According to Herzog, most of the world has become embarrassed. And his is a kind of lone quest against this fact.' Derek Malcolm, [interview with Werner Herzog], *The Guardian*, 24 November 1975. See also Rost (ed.), *Werner Herzog in Bamberg*, 142ff.

[24.] See Peter W. Jansen, 'innen/außen/innen. Funktionen von Landschaft und Raum', in *Herzog/Kluge/Straub. Reihe Film 9*, 69–84, here 69.

[25.] Ibid., 70.

[26.] Hauff's tale and related stories (by Tieck, Schubert, Arnim, E. T. A. Hoffmann, Wagner and others) are anthologized, with a critical essay exploring the theme of the 'cold heart' as a vehicle of critique of industrial society in Romantic and other nineteenth-century literature, in Manfred Frank (ed.), *Das kalte Herz. Texte der Romantik* (Frankfurt am Main, 1978).

[27.] Interview with Chris Peachment, *The Independent on Sunday*, 18 August 1991, 13. See also Rost (ed.), *Werner Herzog in Bamberg*, 143, where Herzog expands on identical comments as recorded in Les Blank's

Burden of Dreams.

28. See Kent Casper, 'Herzog's quotidian apocalypse: *La Soufrière*', *Film Criticism* 15 (1991), 29–37.

29. Peucker in Phillips (ed.), *New German Filmmakers*, 168.

30. For example: 'These [elemental] landscapes are extreme ends of what our planet is all about. And they have enormous visual force. You can stylize and direct the desert and the jungle. Both of them are very good characters, and you can modify them as you would human characters.' Interview by Jonathan Gott, 'Jungle madness', *Rolling Stone*, 11 November 1982, 21.

31. Discussion with Werner Herzog at Longford Cinema, Melbourne, 12 February 1982, transcribed in an untitled pamphlet published by the Australian Film Institute, 2. Almost verbatim in Gene Walsh (ed.), *Images at the Horizon: A Workshop with Werner Herzog conducted by Roger Ebert at the Facets Multimedia center, Chicago, Illinois, April 17, 1979* (Chicago, 1979), 21.

32. Herzog recounts the journey in *Vom Gehen im Eis. München–Paris 23.11 bis 14.12.1974* (Munich 1978). For disdainful commentary, see Jan-Christopher Horak, 'W.H. or the mysteries of walking in ice', in Corrigan (ed.), *Between Mirage and History*, 23–44.

33. Kraft Wetzel, 'Kommentierte Filmographie', in *Werner Herzog. Reihe Film 22*, 87–144, here 123.

34. Rentschler in Corrigan (ed.), *Between Mirage and History*, 176.

35. Ibid., 174.

36. The parallel is explored in detail in Gerhard Kaiser's essay, published in the series Themen – Eine Privatdruck-Reihe der Carl Friedrich von Siemens Stiftung: *Fitzcarraldo Faust – Werner Herzogs Film als postmoderne Variation eines Leitthemas der Moderne* (Munich, 1993).

37. Davidson, 'As others put plays upon the stage', 129–30.

38. Jean Baudrillard, 'The reality gulf', *The Guardian*, 11 January 1991, 25; *La Guerre du Golfe n'a pas eu lieu* (Paris, 1991); and cf. *L'Illusion de la fin* (Paris, 1992), tr. Chris Turner, *The Illusion of the End* (Stanford, 1994); and Paul Virilio, *L'Écran du désert* (Paris, 1991), tr. Bernd Wilczek, *Krieg und Fernsehen* (Munich, 1993). A critique: Chris Norris, *Uncritical Theory: Postmodernism, Intellectuals and the Gulf War* (London, 1992); a defence: Benno Wagner, 'Normality – exception – counter-knowledge', in Mike Featherstone et al. (eds.)., *Global Modernities* (London, 1995), 178–91, here 188. See also, for a different perspective: Marie Gillespie, *Television, Ethnicity and Cultural Change* (London and New York, 1995), 131–41.

39. Werner Herzog, 'Globetrotter with a dream', *The Economist*, 20 October 1984, 104.

40. Davidson, 'As others put plays upon the stage', 129.

41. Ibid., 130.

42. Jacques Derrida, 'Of an apocalyptic tone recently adopted in philosophy', tr. John Leavey, Jr., *The Oxford Literary Review*, 6, 2 (1984), 3–37, esp. 20f. For commentary, see Goldsmith, *Unbuilding Jerusalem*, 1–26, esp.14–6; and cf. Derrida, 'No apocalypse, not now', tr. Catherine Porter and Philip Lewis, *Diacritics*, 14, 2 (1984), 20–31.

43. Hartmut Böhme, 'Vergangenheit und Gegenwart der Apokalypse', in *Natur und Subjekt* (Frankfurt am Main, 1988), 380–98, here 388.

44. Ibid., 392.

45. Ibid., 389.

46. Ibid., 383.

47. Ibid., 393-4. (Schiller's closely related theorizing of the sublime and apocalypse is also discussed in Goldsmith, *Unbuilding Jerusalem*, 6f.)

48. Ibid., 394.

49. Ibid., 396.

50. Ibid., 396-7. The modern literary *locus classicus* of this vision is the closing speech by Möbius in Friedrich Dürrenmatt's *Die Physiker* (1962).

51. Amos Vogel, 'Lessons in darkness', 70.

52. Herzog in Walsh (ed.), *Images at the Horizon*, 22.

53. See Simon Schama, '*Der Holzweg*: the track through the woods', in *Landscape and Memory* (London, 1995), 75–134.

54. Amos Vogel, 'Lessons in darkness', 70.

55. On the making of this film, see Christian Jennings, 'As Kuwait burned', *The Sunday Times*, 23 February 1992, 11 (article based on interviews with Herzog, his cameramen and his helicopter pilot).

56. Koepnick, 'Colonial forestry', 159.

Select Bibliography

Abelshauser, Werner (ed.), *Umweltgeschichte. Umweltverträgliches Wirtschaften in historischer Perspektive* (Göttingen, 1994)

Adorno, Theodor W., *Negative Dialectics* (London, 1973)

Adorno, Theodor W. and Horkheimer, Max, *Dialectic of Enlightenment* (London, New York, 1979)

Bahro, Rudolf, *From Red to Green* (London, 1984)

—— *Radikalität im Heiligenschein. Zur Wiederentdeckung der Spiritualität in der modernen Gesellschaft* (Berlin, 1984)

—— *Logik der Rettung. Wer kann die Apokalypse aufhalten?* (Stuttgart and Vienna, 1989)

Beck, Ulrich (ed.), *Politik in der Risikogesellschaft. Auf dem Weg in eine andere Moderne* (Frankfurt am Main, 1991)

—— *Die Erfindung des Politischen* (Frankfurt am Main, 1993)

Bergmann, Klaus, *Agrarromantik und Großstadtfeindschaft* (Meisenheim am Glan, 1970)

Betz, Hans-Georg, 'The post-modern challenge: from Marx to Nietzsche in the West German alternative and green movement', *History of European Ideas*, 11 (1989), 815–30

Blühdorn, Ingolfur, Krause, Frank, and Scharf, Thomas (eds.), *The Green Agenda: Environmental Politics and Policy in Germany* (Keele, 1995)

Böhme, Gernot, *Für eine ökologische Naturästhetik* (Frankfurt am Main, 1989)

—— *Natürlich Natur. Über Natur im Zeitalter ihrer technischen Reproduzierbarkeit* (Frankfurt am Main, 1992)

Böhme, Hartmut, *Natur und Subjekt. Versuch zur Geschichte der Verdrängung* (Frankfurt am Main, 1988)

—— 'Aussichten einer ästhetischen Theorie der Natur', in Jörg Huber (ed.), *Wahrnehmung von Gegenwart. Interventionen* (Basle and Frankfurt am Main, 1992), 31–53

Bramwell, Anna, 'Ricardo Walther Darré – was this man "Father of the Greens"?', *History Today*, 34 (1984), 7–13

—— *Blood and Soil: Walther Darré and Hitler's 'Green Party'* (Abbotsbrook, 1985)

—— 'A green land far away: a look at the origins of the green movement',

Journal of *the Anthropological Society of Oxford*, 17 (1986), 191–206

Bramwell, Anna, *Ecology in the Twentieth Century: A History* (New Haven and London, 1989)

—— *The Fading of the Greens: The Decline of Environmental Politics in the West* (New Haven and London, 1994)

Brockmann, Stephen, 'The green battlefield: aesthetics and "Kulturpolitik" in the current debate', *German Life and Letters*, 43, 3 (1990), 280–9

Brüggemeier, Franz-Josef and Rommelspacher, Thomas (eds.), *Besiegte Natur. Geschichte der Umwelt im 19. und 20. Jahrhundert* (Munich, 1989)

Bullivant, Keith and Ridley, Hugh (eds.), *Industrie und deutsche Literatur. Eine Anthologie* (Munich, 1976)

Caldecott, Leonie and Leland, Stephanie, *Reclaim the Earth: Women Speak Out for Life on Earth* (London, 1983)

Clapp, B. W., *An Environmental History of Britain since the Industrial Revolution* (London and New York, 1994)

Dalton, Russell J., *The Green Rainbow: Environmental Groups in Western Europe* (New Haven and London, 1994)

Derrida, Jacques, 'Of an apocalyptic tone recently adopted in philosophy', tr. John Leavey, Jr., *The Oxford Literary Review*, 6, 2 (1984), 3–37

—— 'No apocalypse, not now', tr. Catherine Porter and Philip Lewis, *Diacritics*, 14, 2 (1984), 20–31

Diamond, Irene and Orenstein, Gloria (eds.), *Reweaving the World: The Emergence of Ecofeminism* (San Francisco, 1990)

Dittmers, Manuel, *The Green Party in West Germany: Who are they? And What do they Really Want?* (Buckingham, 1988)

Dobson, Andrew, *Green Political Thought*, 2nd edn. (London, 1995)

Dobson, Andrew and Lucardie, Paul (eds.), *The Politics of Nature: Explorations in Green Political Theory* (London and New York, 1993)

Dominick, Raymond H., *The Environmental Movement in Germany: Prophets and Pioneers 1871–1971* (Bloomington and Indianapolis, 1992)

Dowlby, David, *Fictions of Nuclear Disaster* (Houndsmill and London, 1987)

Eckersley, Robyn, *Environmentalism and Political Theory: Towards an Ecocentric Approach* (London, 1992)

Engels, Friedrich, *Dialektik der Natur* (1871), in Marx/Engels, *Werke* 20 (Berlin (DDR), 1956ff.), 307–572

Enzensberger, Hans Magnus, 'Ökologie und Politik oder Die Zukunft der Industrialisierung', *Kursbuch 33* (1973), 1–42. Translated as 'A critique of political ecology', in *Dreamers of the Absolute: Essays on Politics, Crime and Culture* (London, 1988), 253–95

Evans, Richard, *Death in Hamburg: Society and Politics in the Cholera Years, 1830–1910* (Oxford, 1987)

Gaard, Greta (ed.), *Ecofeminism: Women, Animals, Nature* (Philadelphia, 1993)

Gasman, Daniel, *The Scientific Origins of National Socialism: Social Darwinism in Ernst Haeckel and the German Monist League* (London and New York, 1971)

Goldsmith, Edward, *The Great U-Turn: De-industrializing Society* (Bideford, 1988)

Goldsmith, Steven, *Unbuilding Jerusalem: Apocalypse and Romantic Imagination* (Ithaca and London, 1993)

Goodbody, Axel, *Natursprache. Ein Konzept der Romantik in der modernen Naturlyrik* (Kiel, 1984)

—— '"Es stirbt das Land an seinen Zwecken": Writers, the environment and the green movement in the GDR', *German Life and Letters*, 47, 3 (1994), 325–36

Goodin, Robert E., *Green Political Theory* (Cambridge, 1992)

Gould, Peter C., *Early Green Politics: Back to Nature, Back to the Land, and Socialism in Britain 1880–1900* (Brighton and New York, 1988)

Green, Michael, *Mountain of Truth: The Counterculture Begins. Ascona, 1900–1920* (Hanover, NH, and London, 1986)

Grimm, Gunter E., Faulstich, Werner and Kuon, Peter (eds.), *Apokalypse. Weltuntergangsvisionen in der Literatur des 20. Jahrhunderts* (Frankfurt am Main, 1986)

Grimm, Reinhold, '"The ice age cometh": a motif in modern German literature', in Siegfried Mews (ed.), *'The Fisherman and His Wife': Günter Grass's 'The Flounder' in Critical Perspective* (New York, 1983), 1–17

Grimm, Reinhold and Hermand, Jost (eds.), *From the Greeks to the Greens: Images of a Simple Life* (Madison and London, 1989)

Gsteiger, Manfred, 'Zeitgenössische Schriftsteller im Kampf für die Umwelt', in Manfred Schmeling (ed.), *Funktion und Funktionswandel der Literatur im Geistes- und Gesellschaftsleben. Akten des Internationalen Symposiums Saarbrücken 1987* (Berne and Frankfurt am Main, 1989)

Hahn, Gerhard, *Die Freiheit der Philosophie. Eine Fundamentalkritik der Anthroposophie* (Göttingen, 1995)

Haraway, Donna, *Simians, Cyborgs, and Women: The Reinvention of Nature* (New York, 1991)

Hayward, Tim, *Ecological Thought: An Introduction* (Cambridge, 1994)

Hermand, Jost, *Grüne Utopien in Deutschland. Zur Geschichte des ökologischen Bewußtseins* (Frankfurt am Main, 1991)

—— (ed.), *Mit den Bäumen sterben die Menschen. Zur Kulturgeschichte der Ökologie* (Cologne, Weimar and Vienna, 1993)

Hermand, Jost and Müller, Hubert (eds.), *Öko-Kunst. Zur Ästhetik der Grünen* (Hamburg, 1989)

Hope, Jacquie, *Green Trends in East Germany: Critiques of Modern Industrial Society in GDR Literature*, D.Phil. thesis, Oxford, 1992

Horkheimer, Max, *Eclipse of Reason* (New York, 1947)

Hülsberg, Werner, *The German Greens: A Social and Political Profile* (London, 1988)

Jefferies, Matthew, *Politics and Culture in Wilhelmine Germany: The Case of Industrial Architecture* (Oxford, 1995)

Jordan, Peter, 'Carl Hiaasen's environmental thrillers: crime fiction in search of green peace', *Studies in Popular Culture*, 13, 1 (1990), 61–71

Kelly, Alfred, *The Descent of Darwin: The Popularization of Darwinism in Germany, 1860–1914* (Chapel Hill, 1981)

Klueting, Edeltraud (ed.), *Antimodernismus und Reform. Zur Geschichte der deutschen Heimatbewegung* (Darmstadt, 1991)

Knabe, Hubertus, 'Zweifel an der Industriegesellschaft. Ökologische Kritik in der erzählenden DDR-Literatur', in Werner Gruhn et al., *Umweltprobleme und Umweltbewußtsein in der DDR* (Cologne, 1985), 201–50

Kolinsky, Eva (ed.), *The Greens in West Germany: Organisation and Policy Making* (Oxford, 1989)

Kroeber, Karl, *Ecological Literary Criticism. Romantic Imagining and the Biology of Mind* (New York, 1994)

Kurz, Paul Konrad, *Apokalyptische Zeit. Zur Literatur der mittleren 80er Jahre* (Frankfurt am Main, 1987)

Lilienthal, Volker, 'Irrlichter aus dem Dunkel der Zukunft. Zur neueren deutschen Katastrophenliteratur', in Helmut Kreuzer (ed.), *Pluralismus und Postmodernismus. Zur Literatur- und Kulturgeschichte der achtziger Jahre*, 2nd edn. (Frankfurt am Main, 1991), 190–224.

Link, Thomas, *Zum Begriff der Natur in der Gesellschaftstheorie Theodor W. Adornos* (Cologne and Vienna, 1986)

Linse, Ulrich (ed.), *Zurück, o Mensch, zur Mutter Erde. Landkommunen in Deutschland, 1890–1933* (Munich, 1983)

—— *Ökopax und Anarchie. Eine Geschichte der ökologischen Bewegungen in Deutschland* (Munich, 1986)

McKibben, Bill, *The End of Nature* (New York, 1989)

Mayer-Tasch, Peter C., *Natur denken. Eine Genealogie der ökologischen Idee* (Frankfurt am Main, 1991)

Merchant, Carolyn, *The Death of Nature: Women, Ecology and the Scientific Revolution* (New York, 1980)

—— *Radical Ecologies: The Search for a Livable World* (London and New York, 1992)

Meyer, E.Y., *Plädoyer – Für die Erhaltung der Vielfalt der Natur beziehungsweise für deren Verteidigung gegen die ihr drohende Vernichtung durch die Einfalt des Menschen* (Frankfurt am Main, 1982)

Muschg, Adolf, *Empörung durch Landschaften. Vernünftige Drohreden* (Frankfurt am Main, 1988)

Muthesius, Stefan, 'The origin of the German conservation movement', in Roger Kain (ed.), *Planning for Conservation* (London, 1981)

Otto, Christian F., 'Modern environment and historical continuity', *Art Journal* (1983), 148–57

Papadakis, Elim, *The Green Movement in West Germany* (London, 1984)

Pepper, David, *The Roots of Modern Environmentalism* (London, 1986)

—— *Eco-socialism: From Deep Ecology to Social Justice* (London and New York, 1993)

—— *Modern Environmentalism: An Introduction* (London, 1996)

Plant, Judith (ed.), *Healing the Wounds: The Promise of Ecofeminism* (Philadelphia, 1989)

Ponting, Clive, *A Green History of the World* (London, 1991)

Princen, Thomas and Finger, Matthias, *Environmental NGOs in World Politics* (London, 1994)

Raschke, Joachim, *Die Grünen. Wie sie wurden, was sie sind* (Cologne, 1993)

Richardson, Dick and Rootes, Chris (eds.), *The Green Challenge: The Development of Green Parties in Europe* (London and New York, 1995)

Rucht, Dieter, 'Environmental movement organisations in West Germany and France: Structure and interorganizational relations', *International Social Movement Research*, 2 (1990), 61–94

Schama, Simon, *Landscape and Memory* (London, 1995)

Scharf, Thomas, *The German Greens: Challenging the Consensus* (Oxford and Providence, 1994)

Seager, Joni, *Earth Follies: Feminism, Politics and the Environment* (London, 1993)

Short, John Rennie, *Imagined Country: Environment, Culture and Society* (London, 1991)

Sieferle, Rolf Peter, *Fortschrittsfeinde? Opposition gegen Technik und Industrie von der Romantik bis zur Gegenwart* (Munich, 1984)

Spretnak, Charlene and Capra, Fritjof, *Green Politics: The Global Promise* (London, 1984)

Trepl, Ludwig, *Geschichte der Ökologie vom 17. Jahrhundert bis zur Gegenwart* (Frankfurt am Main, 1987)

Vondung, Klaus, *Die Apokalypse in Deutschland* (Munich, 1988)

Wall, Derek, *Green History: A Reader in Environmental Literature, Philosophy and Politics* (London and New York, 1994)

Warren, Karen (ed.), *Ecological Feminism* (London and New York, 1994)

Weinzierl, Hubert, *Das grüne Gewissen. Selbstverständnis und Strategien des Naturschutzes* (Stuttgart and Vienna, 1993)

Wiesenthal, Helmut, *Realism in Green Politics: Social Movements and*

Ecological Reform in Germany, ed. John Ferris (Manchester, 1993)

Worster, Donald, *Nature's Economy: A History of Ecological Ideas*, 2nd edn. (Cambridge, 1994)

Zimmermann, Jörg (ed.), *Das Naturbild des Menschen* (Munich, 1982)

Index